The Chin

The Chin

Edited by

J. Lévignac

Director of Post-Graduate Education, European Association
for Cranio-maxillofacial Surgery; Attending Surgeon,
Department of ENT and Maxillofacial Surgery,
Hôpital Lariboisière, Paris, France

Translated by

S. Anthony Wolfe MD FACS

Clinical Professor of Plastic and Reconstructive Surgery,
University of Miami School of Medicine, Miami, Florida, USA

CHURCHILL LIVINGSTONE
EDINBURGH LONDON MELBOURNE AND NEW YORK 1990

CHURCHILL LIVINGSTONE
Medical Division of Longman Group UK Limited

Distributed in the United States of America by
Churchill Livingstone Inc., 1560 Broadway, New York,
N.Y. 10036, and by associated companies, branches
and representatives throughout the world.

First edition 1990

ISBN 0 443 042217

British Library Cataloguing in Publication Data
The Chin.
1. Man. Mandible. Plastic surgery
I. Levignac, J.
617′.522

Library of Congress Cataloging in Publication Data
Menton. English.
   The Chin/edited by J. Lévignac, with the collaboration
      of G. Aiach . . . [et al.]; translated by S. Anthony
   Wolfe. — 1st ed.
      p.   cm.
   Translation of: Le Menton.
   Includes index.
   1. Chin — Surgery.   2. Surgery, Plastic. I. Lévignac, J.
II. Title.
      [DNLM: 1. Chin — surgery.   WE 705 M549]
   RD526.5.M4613   1990
   617.5′22 — dc20
   DNLM/DLC
   for Library of Congress   89-17464
                                                            CIP

Le Menton sous la direction de J. Lévignac
© Masson, Editeur, Paris, 1988

Produced by Longman Singapore Publishers (Pte.) Ltd.
Printed in Singapore.

Contributors

G. Aiach
Attending Surgeon, Department of Plastic and Maxillofacial Surgery, C. H. U., Créteil, France

J. L. Cariou
Chief, Department of Stomatology and Maxillofacial Surgery, I. H. A. Jean Prince, Papeete, Tahiti

J. C. Chalaye
Chief, Department of Maxillofacial Surgery, Hôpital de Saintes, Saintes, France

J. Dautrey
Maxillofacial Surgeon, Clinique St. Andre, Vandoeuvre les Nancy, France

J. Delaire
Chief, Department of Stomatology and Maxillofacial Surgery, C. H. R., Nantes, France

G. Despreaux
Attending Surgeon, ENT Department, Hôpital Avicenne, Bobigny, (Seine-St-Denis), France

P. A. Diner
Resident Surgeon, Department of Stomatology and Maxillofacial Surgery, Hôpital Pitié-Salpêtrière, Paris, France

H. P. Freihofer
Chief, Department of Maxillofacial Surgery, Heelkundige Klinieken, Afdeling Mond and Kaakchirurgie, Nijmegen, The Netherlands

D. Ginesty
Chief, Department of Stomatology and Maxillofacial Surgery, Hôpital St-Vincent-de-Paul, Paris, France

J. L. Heim
Department of Anthropology, Musée de l'Homme, Paris, France

J. Lévignac
Director of Post-Graduate Education, European Association for Cranio-maxillofacial Surgery; Attending Surgeon, Department of ENT and Maxillofacial Surgery, Hôpital Lariboisière, Paris, France

J. Mercier
Attending Surgeon, Department of Stomatology and Maxillofacial Surgery, C. H. R., Nantes, France

L. Merville
Chief, Department of Maxillofacial and Plastic Surgery, Hôpital Foch, Suresnes, France

P. Oxeda
Resident Surgeon, Department of Stomatology and Maxillofacial Surgery, Hôpital Pitié-Salpêtrière, Paris, France

A. Pasturel
Chief, Department of Stomatology and Plastic Surgery, Hôpital des Armées Bégin, Saint-Mandé, France

A. Petrovic
Director of Research, Inserm, Strasbourg, France

B. Richbourg
Chief, Department of Maxillofacial and Plastic Surgery, C. H. R., Besançon, France

F. Souyris
Chief, Department of Stomatology and Maxillofacial Surgery, Hôpital Lapeyronie, Montpellier, France

J. Stutzmann
Research Department, Inserm, Strasbourg,
France

J. F. Tulasne
Plastic and Maxillofacial Surgeon, Clinique
'Château du Belvédère', Boulogne-sur-Seine,
France

J. M. Vaillant
Chief, Department of Stomatology and
Maxillofacial Surgery, Hôpital Pitié-Salpêtrière,
Paris, France

Contents

1. Introduction

J. Lévignac

The symposium on the 'chin' carried out under the aegis of the French Association of Maxillofacial Surgeons is intended for all those individuals — dermatologists, plastic surgeons, orthodontists and otorhinolaryngologists — who are interested in the forms of the face, who are called upon to modify the face if necessary, and who want to understand any 'anomaly' to better be able to deal with it when required and even, in certain cases, to prevent it.

In fact, the chin, a small object, carries us quite far away if we want to understand its significance. The quadruped mammal does not have a chin; this projection does not appear in animal evolution until the advent of the upright biped posture. The chin is one of the characteristics of Man. How did it come to be, and why?

To answer these questions one needs to study the genesis of the chin and its development, noting here how heredity expresses itself in singularities.

The essential relationship between the biology of bone, its inherent rhythms and the influences which play a part will come from fundamental research.

Since one cannot isolate the chin from the lower jaw as it takes part in masticatory dynamics, one can see how important forces come to play leading to this bony reinforcement as a necessary 'architectural response'.

Seen through human evolution, the projection of the chin is closely linked with the regression of the dental apparatus; this is one of the lessons that can be drawn from anthropology.

All this helps our understanding, but does not fully explain the morphological variations which are seen, and it is here that one recognizes the importance of the muscular environment which is also acting on the surface.

Finally, the chin can be seen as an essential reference point in the overall cervicocephalic profile and this relates it closely to postural phenomena which place the head in equilibrium on the top of the vertebral axis. In discussing this, one might think that one is wandering too far away from the subject of the chin, but this perspective is necessary to better understand the maxillomandibular alterations and appreciate the value of the chin as a reference point in the overall profile.

All this having been noted, and with a complete knowledge of the cause in the analysis of a particular disorder, one can then undertake planning for the ideal correction.

The plan for correction evolves from a complete architectural analysis, which establishes with precision the changes in dimension which will occur and takes account of the equilibrium of the forces which should assure the result. In a whole series of characteristic anomalies, we will see the methods used for correction and the long-term results.

This symposium ends with an important chapter on 'Mutilations of the chin and their treatment'.

Without doubt, in such a work, everything cannot be said and certain questions will remain unanswered, but we feel that it is useful to focus attention on various aspects of the chin in one small monograph. The experience of the others who took part in this symposium is considerable, but it is only possible to deal with the essential points.

We wanted to focus on practical aspects in our discussions. Our aim has been to be useful in both technical and artistic aspects and also to provoke thought and reflection.

We should point out that the work in this monograph came from presentations given at the XII Congress of the French Society of Maxillo-facial Surgeons, under the presidency of J. Dautrey (Nancy, France, 1985).

We would like to thank A. Pasturel and G. Aiach for their important collaboration in putting the book together. We would also like to particularly express our gratitude to Mr and Mrs Milton Cassel for their great interest in the French Society of Maxillofacial Surgeons and for their generous contribution.

2. Morphogenesis of the human chin

D. Ginisty

INTRODUCTION

The aim of this work is to try to understand the morphogenesis of the mental trigone in contemporary man. In effect, as was emphasized by Piveteau (1963) who cited Weidenreich (1936):

The genesis of the projection of the chin results from the action of two processes: the individualization of a mental triangle (*trigonum mentale*) and the formation of a depression (anterior *incurvatio mandibularis*) at the junction of the alveolar portion and the basilar portion of the mandible, these two components being independent of each other.

The study of the genesis of the projection of the human chin finds an application in paleoanthropology and in the study of primatology because the chin is a morphological characteristic of *homo sapiens*, both in comparison with fossilized specimens and in comparison with other primates.

The formation of the anterior *incurvatio mandibularis* is explained by the regression of the alveolar ridge which supports the teeth on the basilar portion, a regression which becomes accentuated during growth (Du Brul 1954); the thinness of the dental system on the masticatory apparatus associated with the need to preserve upper airways with the upright biped posture are the determining aetiological elements.

The individualization of the mental trigone is the second process in the genesis of the mental projection.

Weidenreich (1936) proposed a phyletic interpretation of the evolution of the chin region in hominids and the formation of the mental trigone in *homo sapiens*. According to him, originally the two dental bones formed a 'V' in the horizontal plane; the bony dental margins were in close contact along their entire height. In the hominids,

little by little there was frontalization of the anterior portion of the two dental bones, which rested in contact only at the alveolar level: as a consequence, there occurred a separation of the two dental bones in their basilar portion. This open space would be filled in by the formation of ossicles.

The main critiques of this hypothesis come from examining fossil mandibles and mandibles of large, present-day monkeys. In fact, the fossil mandibles of hominids or prehuman humanoids never have a basilar arch form in the shape of a 'V' and they do not have the elements of the mental trigone.

Furthermore, one does not find this type of 'V'-shape among the great apes of today: the two dental bones form a 'U' with a complete frontalization of the horizontal branches in the incisive regions; the dental bones are completely in contact along the entire height of the symphyseal region, and without any known type of ossicle having been described.

The hypothesis of Weidenreich is, therefore, not convincing. It has always seemed important to us to study the ontogenesis of the mental region in man. This takes place during the first 18 months of postnatal life, at a time when the basilar bone and the alveolar bone are not yet defined. In effect, at birth the two dental bones, still separated, are quasi-tubilar osseous structures surrounding the dental germs which will constitute the deciduous dentition.

The change from a double structure to a single structure takes place in man, always after the appearance of the ossicles, of varying number, which fill the interosseous dental space before joining the two hemimandibles together and growing on to form the mental symphesis. These

3

ossicles, which have not been observed in primates other than man, are the direct origin of the mental trigone in the newborn.

However, with the formation of the *trigonum mentale* and the process of the formation of the symphysis, the morphological difference between man and other primates becomes accentuated.

Our work has analyzed the histological phenomenon of ossification of the chin in the peri- and postnatal periods by focusing on the intermediate stages of the formation of the symphysis, dominated in large part by the presence of ossicles in the interosseous dental space.

The term interosseous dental space (IODS) defines the anatomical hiatus that separates the two dental frameworks before the process of symphysation (formation of the symphysis).

MATERIALS AND METHODS

Materials

This work was carried out on the following specimens:

1. seven newborns on whom we were not able to obtain any information as to length of gestation or cause of death;
2. a two-month old infant;
3. a five-month old infant;
4. an infant of approximately 15 months of age.

All of these infants were taken to be normal. Where possible, cranial circumference was measured. Each chin specimen was photographed on a dental film before decalcification which made it possible to determine the width of the interosseous dental space, the presence or absence of ossicles and the degree of mineralization of the buds of the deciduous incisive and canine teeth (Fig. 2.1).

Methods

The study was carried out on semiserial microscopic sections. Each mental region was taken as a block including the skin, the lower lip, the tongue and the hyoid in certain cases. The specimen was decalcified and then mounted in

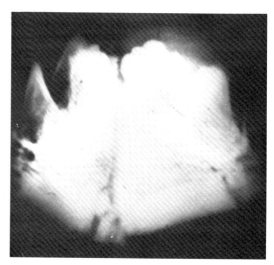

Fig. 2.1 Radiograph of a chin specimen. The ossicles are quite visible between the two dental bones.

paraffin. Sections 5 to 7 μm thick were taken after every tenth cut, which would be approximately every 50 to 70 μm. Finally, the sections were stained either by a hematoxylin–eosin or by a Masson trichrome stain.

The sections were taken in the three different planes of space: frontal and sagittal for the two newborn specimens; horizontal for all the others.

Results

We will first describe the IODS and its anatomical relationships before going on with the study of the orientation of the connective tissue fibres and the distribution of the cartilaginous tissues.

The interosseous dental space

1. The IODS in the perinatal period before the appearance of the ossicles. This can be described by studying the chin of newborns who do not show any radiologically detectable ossicles.

a. The interosseous space: its topographical complexity (Fig. 2.2). The mesial extremities of each dental bone are separated by a space of varying width depending on its level. In the upper half of the space, the two edges are parallel but the space is not always rectilinear; according to the cuts and the specimens, it is more or less sinusoid. There

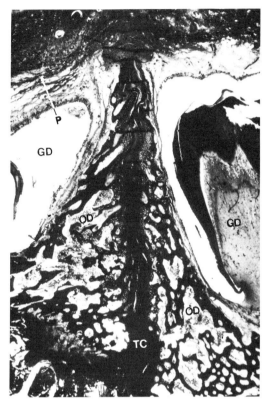

Fig. 2.2 Horizontal section of the chin of a newborn. GD = dental germ; OD = dental bone; P = periosteum; TC = intermediate connective tissue.

are digitations which cause variations in its width by a mean of 0.5 mm.

In the lower half of the space, the two edges are no longer parallel. The space is triangular with an inferior base and a superior apex on frontal cuts. The maximum width is approximately 3.5 to 4 mm from the inferior pole of the space, both anteriorly and posteriorly.

The obliquity of the edges varies considerably and is not symmetrical if one compares the right and left dental bones as well at the anterior and posterior portions of the space. This explains why the space can, on certain horizontal cuts, have the aspect of an asymmetrical rhomboid.

b. The structure of the IODS. This space is occupied by fibrous connective tissue in complete continuity with the periosteum of the mandible at all levels and on the surfaces of the mandible. It has the same morphological structure as periosteal tissue and is richly vascularized through all

sections on its inner surface. Just as the periosteum follows the contours of the two hemimandibles, the connective tissue closely follows the surface of the bony edges and has digitations like the fingers of a glove which spread in different directions.

At this level there is a kind of cellular sleeve around each of these osseous edges — a cartilaginous sleeve in complete continuity both with the connective tissue and with the bone. This cartilaginous sleeve has a polymorphic appearance (Fig. 2.3) with:

1. a zone of elongated cells of the fibroblast type on the medial surface, forming a dense layer in the connective tissue;
2. a distal, osseous side, where there is a zone of hypertrophic cartilaginous cells in contact with vascular buds located in the mandibular bone;
3. a layer of cells between the two zones which are rounder than the fibroblasts, surrounded by a chondroid matrix.

Examination under high-powered magnification makes it possible to confirm the great coherence of all the structures described above – the connective tissue layer, the cartilaginous layer and the bone layer which provide evidence of a zone of ossification and, therefore, functional growth.

2. The stage of the ossicles. Two, three or four ossicles occupy the interdental space which was previously described. When there are four, one finds two inferior and anterior ossicles and two superior and posterior ossicles. On the microscopic sections, one notes that they are located in dense, fibrous, connective tissue separating the two dental bones and have, on their periphery, a cartilaginous border identical to that covering the osseous framework. One observes that at this level there is a great cohesiveness with the dense, connective tissue since one can move directly, without discontinuity, from the fibrous, connective tissue which is quite vascularized to the connective tissue which is rich in fibroblasts, on to the evolving stages of the cartilaginous cells and on again to the bone (which itself is centrally located).

a. Relationships of the ossicles with the mandibular framework. One can see that the radiological autonomy of the ossicles, compared to

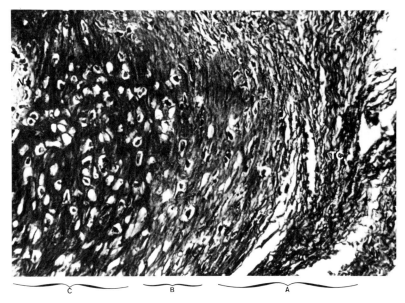

Fig. 2.3 Horizontal section of the chin of a newborn. A. = fibrous layer; B = compartment of growth of intermediate secondary cartilage, zone of skelettoblasts and prechondroblasts; C = zone of chondroblasts; TC = intermediate connective tissue; V = vessels.

the ipsilateral bone, does not exist in an absolute fashion on the microscopic sections.

When sections have been made across the entire upper portion of the zone of the ossicles it may be seen that there is always a bridge or union between the framework and the adjacent ossicle.

Depending upon the level examined one can note either an ossicle which is completely isolated, in which case the peripheral cartilaginous layer is much more dense than the mandibular framework, or an ossicle which is partly integrated into the skeletal framework. In the latter case, the cartilaginous layer is more dense than the contralateral ossicle.

b. Relationship between the ossicles themselves. As described above, the ossicles are surrounded by connective tissue except in the areas where they are in contact with the homolateral hemimandible. At this stage they are never connected to one another across the midline. To the contrary, one can see in areas where two ossicles are superimposed that there are zones of cartilaginous tissue with hypertrophic cells situated directly between the osseous tissues and the two homolateral ossicles. This produces a chondrolysis

which precedes the stages of mineralization and then ossification.

3. Relationships of the interosseous dental space. As for the space between the two hemimandibles, whether or not they are individual ossicles, the anatomical relationships with the surrounding muscular-periosteal elements are identical. We will successively examine the posterior, inferior, anterior and superior relationships of this particular space (Fig. 2.4).

a. Posterior relationships of the interosseous space. The posterior surface of the sutural zone contains the insertion of the genioglossus muscles. These muscles have a direct insertion on the interosseous connective tissue, at the point where the space is widest. There is no continuity between the median lingual raphé and the connective tissue.

As to the insertions of the geniohyoid muscles, they are on the portion just next to each mandibular border, this being at a more inferior level than the area of insertion of the genioglossus muscles. In the zone of transition from the insertion of the genioglossus muscles to the geniohyoid muscles it is difficult to distinguish them, and one cannot establish that there are no insertions of the

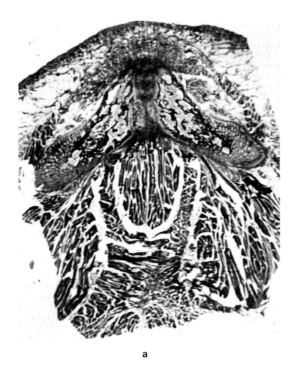

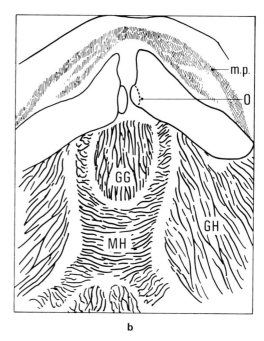

a   b

Fig. 2.4 Horizontal section (specimen X_1). O = ossicle; GC = genioglossus muscle; MH = mylohyoid muscle; GH = geniohyoid muscle; MP = platysma muscle.

geniohyoid muscles under the connective tissue of the interosseous space.

b. Posterior surface: digastric muscles. The digastric muscles, in their terminal portion, are composed of tendinous fibres which insert on the interosseous space. On horizontal sections taken at two months one can examine the insertions of these muscles. With certain microscopic techniques, it can be established that these insertions do not take place on the dental-osseous framework but under the anterior and inferior ossicles. On the other horizontal cuts, the absence of the insertion of the digastric muscles into each osseous framework in the area of the interdental space cannot be established.

c. Anterior relationships: the symphysis and cutaneous muscles of the chin. The triangular muscles of the lip and the quadratus muscles of the chin insert under the mandibular periosteum laterally. There is, therefore, no insertion of the skin muscle directly under the connective tissue of the symphysis. In the superior portion of the space, the interdental connective tissue is separated from the muscular planes of the

orbicularis by a richly vascularized, areolar tissue and occupies a triangular space at the apex of the lip in horizontal sections. In the inferior portion of the space, the muscular plane is directly in contact with the periosteal plane. But at both levels, one does not find the insertion of the platysma muscle directly into the connective tissue space.

d. Superior relationships. The connective tissue of the interdental space continues up to the level of the future alveolar crest, in the area of the insertions of the frenulum of the lower lip. One can show at this level that there are muscular fibres of the obicularis passing into the frenulum which extend up to the dental crest.

There, also exist nerve receptors of the Water-Pacini type in the fibromucosa covering this crest.

4. The stage of incomplete synostosis. Horizontal sections taken from the chin of an infant of five months of age with a peripheral cranial circumference of approximately 37.5 cm were studied.

Radiologically, one could see two inferior ossicles attached at the same time to each hemimandible and to each other, as well as a narrow interosseous space above the zone of the ossicles. In the superior half of the space the two osseous borders were more or less connected.

Histologically, one could not find an interdental space except at the level of the superior and inferior sections. Except at these extremities, there existed throughout a bony bridge which united the two skeletal frameworks. This zone of synostosis was located in a variable position depending on the level. It was situated in the medial third of the sections, above the insertions of the genioglossus muscles, and then increasingly towards the posterior as the sections went lower.

5. The stage of total synostosis. An infant of 15 months of age with a cranial circumference of 41.5 cm was studied in horizontal sections. The synostosis was complete except:

a. at the two superior and inferior extremities of the space where the two skeletal frameworks remained separate;
b. at the level of the insertion of the geniohyoid muscles which occurred laterally on the osseous frameworks.

At this age one cannot find a sleeve-like cartilaginous border at the level of the edges of the bone, nor can one find any cartilaginous nodules lying free in the connective tissue.

In sum, the zone of growth described in the newborn had completely disappeared. In the sections where there was a continuity between the two dental bones, there was no longer a zone of enchondral ossification. The connective tissue itself was poor in cells. Also, the mandibular framework was now unique. Ossification took place at the level of the insertions of the genioglossus and the geniohyoid muscles onto the posterior surface of the bone: the newly formed collagen fibres of the bone were in continuity with the tendinous fibres in the insertion of the muscles. Also, each dental bone was made of haversian bone. The fibrous bone which predominated in all the early specimens no longer existed in this stage except at the level of the posterior muscular insertions.

1. Study of the orientation of collagen fibres before symphysation (Figs. 2.5 and 2.6). The study of the orientation of collagen fibres in three

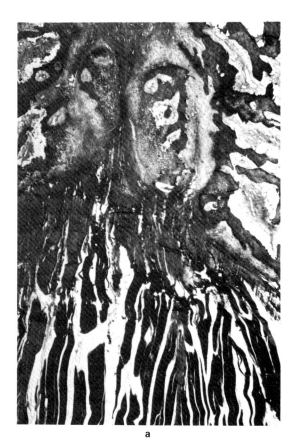

a

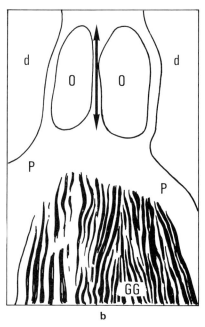

b

Fig. 2.5 Details of the insertions of the genioglossus muscles (specimen X1, horizontal section). O = ossicle; D = dental bone; P = periosteum; GG = genioglossus muscle.

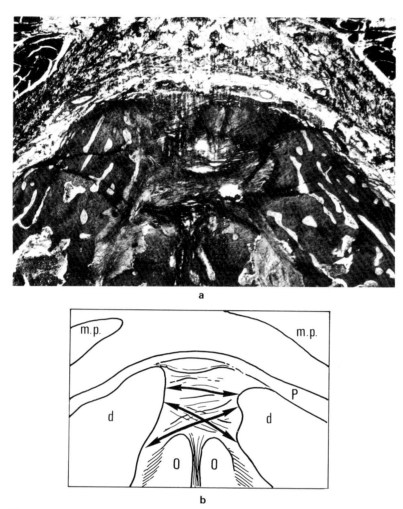

Fig. 2.6 Detail of horizontal section (specimen X1X). Orientation of the fibres of the connective tissue in front of the ossicles. O = ossicle; D = dental bone; P = periosteum; MP = platysma muscle.

series of horizontal sections before the stage of synostosis showed that:

a. In the posterior portion of the connective space (posterior third or half), at the level of the insertions of the genioglossis muscles, the fibres are oriented in a sagittal direction, delimiting one or two ossicles depending on the specimen and the level of the section. They are, therefore, oriented in the direction of the tendinous insertions of these muscles and follow them closely without being in continuity.

Below the insertions of the genioglossus, they are oriented transversally in a dense layer going from one dental bone to the other, and in particular at the level of the insertion to the geniohyoids where they are isolated. One can also see that the tendinous fibres of the insertions of the geniohyoids are oriented posteriorly and laterally (whereas those of the genioglossus seen from above are strictly sagittal).

Above the insertions of the genioglossus, the collagen fibres are oriented in a sagittal plane, but one can also find a very dense periosteum forming a bridge from one mandibular edge to the other.

b. At the anterior pole (anterior third or half), at all levels the periosteum forms a transverse bridge from one mandibular edge to the other. In the posterior portion of the space, well above the ossicles, the fibres are oriented in a sagittal direction as well as posteriorly. At the level of the ossicles, where the interdental space is large, the fibres are always cruciform, between the anterior ossicles and just in front of them, and are transversally oriented further forward with contact with the periosteal fibres.

Depending on the surface of the section containing the ossicles, the quantity of cruciform fibres varies. If the surface is large, the posterior sagittal fibres occupy more than half the space; if the surface is small, which is the case when one sees the superior sections containing the ossicles, the bundles of sagittal fibres occupy the posterior third of the space, cruciform fibres, the middle half or third of the space, and transverse fibres, the anterior third or less.

2. Study of the distribution of the cartilaginous tissue in the perinatal period. The distribution differs according to the stage.

a. Before the appearance of the ossicles one notes a cartilaginous border at the level of the two mandibular edges. This exists along the entire length of these edges but is particularly evident at the anterior and posterior extremities of the space as well as at the superior and inferior poles.

b. At the stage of distinctive ossicles the cartilaginous zones are particularly rich:

(i) at the level of the edge of the mandible and the homolateral ossicle of the sections where the ossicle is separated from the osseous edge;

(ii) at the level of the two ossicles on the two mesial surfaces, when the ossicles are in osseous continuity with the hemimandibles.

3. Study of the orientation of the connective tissue fibres and the cartilage at the time of symphysation. It is only at the superior and inferior poles of the mental region that there remains a space still filled with connective tissue where the fibres are oriented in a sagittal direction.

The synostosis which is median above the insertion of the genioglossus muscles is increasingly situated posteriorly as the sections become more inferior.

This synostosis is not homogeneous. Bridges of cartilaginous union of hypertrophic cells with disseminated blood vessels can be observed. There also exists an anterior cartilaginous bridge at the level of insertion of the triangular muscles of the

lip, associated with a posterior synostosis; between these two bridging areas, the connective tissues are oriented in a sagittal direction.

Higher up, at the level of the posterior insertions of the genioglossus and geniohyoid muscles, one finds in the synostosis (situated at the junction of the medial third and posterior third) small islands of hypertrophic cartilaginous cells.

Still higher, the cartilaginous bridging moves towards the front; there is a cruciform appearance to the connective tissue fibres in the area of the middle third where the synostosis is located. Further to the front, the fibres are oriented sagittally; posteriorly, they are transverse and in continuity with the very dense periosteum.

All through specimen VI, the synostosis appeared to be located differently depending on the level which was examined. At the level of the superior ossicles it was median, above the insertions of the genioglossus muscles; at the level of the inferior ossicles it was located more posteriorly, towards the inferior pole of the chin.

This zone of synostosis, surrounded by cartilage with hypertrophic cells, sometimes containing small islands of hypertrophic cartilaginous cells in contact with vascular buds and osteoclasts, resembles, as in earlier stages, an enchondral ossification, although in its terminal stage.

DISCUSSION

Therefore, at the stage where the dental osseous frameworks are completely separated, or at the stage of the early synostosis, the collagen fibres of the connective tissue of the IODS have an orientation which reflects the traction and shearing forces exerted on them. Numerous authors have studied the role of mechanical factors in the genesis of bone and cartilaginous tissue (Gluckzmann 1939, Murray & Drachman 1969, Basset & Herrmann 1961, Hall 1967, 1970, 1972) and of the orientation of the collagen fibres and connective tissues (Rouvière 1939, Kummer 1963, Flint 1980, Broom 1980, Beecher 1977).

The study of the orientation of the collagen fibres described in the interosseous dental connective tissue show that this zone is under the influence of tension in the following directions:

1. transverse tension, working at all levels and explaining the formation of a solid periosteum; this is particularly important in the anterior half or third of the space, in front of the ossicles. It is associated with shearing tension, responsible for the cruciform orientation of the fibres in front of and between the interior ossicles.
2. sagittal tension, at the level of insertion of the genioglossus muscles in the zone of the posterior ossicles.

We will relate the restraints placed on the interosseous space to the functions of suction and swallowing of the fetus and the newborn, a type of swallowing which is fundamentally different to that of the adult.

1. Chronology of the appearance of these functions

As was emphasized by Anderson (1979), the reflex of suction to buccal stimulation appears at 9 weeks in utero. Swallowing appears at $2\frac{1}{2}$ months, and suction is functional at 3 months at the time when the reactions of flexion and extension of the fetal head occur by contraction of the nuchal muscles.

The fetus sucks its fingers and thumb from the 18th week onwards. Sunction and swallowing are, therefore, very early processes evidencing the neurological activity of the rhombencephalon (Couly 1982); their importance increases after birth. We will briefly review the pertinent muscular physiology.

2. Physiological review

As described by Darque (1972), suction can empty the milk from the mammary passages or from a bottle by two mechanisms:

a. the pressure of the breast or the nipple between the palate and the back of the tongue;
b. breathing.

Both of these mechanisms precede swallowing.

For Robin (1928) suckling begins by a breath during which the end of the breast fills with milk as it is stretched to be placed in the mouth. Here it is seized and placed between the palatal vault and the surface of the tongue which lowers at the same time as the mandible; meanwhile the cheeks contract.

There is also a contraction at the *orbicularis oris*. The second stage of suckling involves the coming together of the jaws which squeeze the breast between them. The third phase begins when the tongue moves progressively from the front to the rear, compressing the breast which is swollen with milk. The *orbicularis oris* tightly contracts in order to assure the seal of the buccal cavity. The tongue and the mandible move upward and forward and together make a posterior movement which immediately precedes the reflex of swallowing (or the pharyngeal portion of swallowing). At this moment the lips open, the mandible moves away from the upper jaw and the tongue is lowered, which permits a new breath.

During the sequences of suction and swallowing, the tongue and mandible have approximately parallel movements, and appear in profile somewhat like the movements of a connecting arm, which alternatively bring into play the elevators (the temporomasseteric complex and medial pterygoid muscle) and then the depressors of the mandible (the lateral pterygoid, the superhyoid muscles and the digastric).

On the other hand, the hyoid (partially ossified, partially cartilaginous) moves very little because the position in the newborn is high (Senecail 1979).

Thus, there is a great difference in the phenomenon of swallowing between the nursing infant and the adult, a difference which involves all the structures that participate in these functions: osteocartilaginous, muscular, and neural.

In sum, the phenomena of suction and swallowing in the fetus and the newborn involve bringing into play in intermittent fashion the muscles cited above. The following phases occur in succession:

1. At the beginning of suction there are contractions of the *orbicularis oris*, all the skin muscles and platysma, as well as the buccinator muscles;
2. A contraction of the elevators of the mandible as well as the skin muscles by which a buccal seal is assured;
3. Finally, a contraction of the entire lingual

musculature which moves from the front to the back, associated with a movement of the mandible by a contraction of its depressors: the superhyoid muscles, anterior belly of the digastrics, mylohyoids and the geniohyoids.

The electromyographic study of swallowing in the newborn carried out by Raimbault et al (1977, 1979) involved the relationships between the three portions of suckling and the pharyngeal portion of swallowing: in the healthy infant the buccal portion of swallowing is characterized by a wave of action potentials in the genioglossus muscles. Their mean duration was 228 ms (168–380 ms). The maximum amplitude was between 300 and 400 μV.

The time period between the beginning of this wave and the beginning of the pharyngeal phase of swallowing, marked by action potentials of the thyrohyoid muscle, is a mean of 455 ms from the onset of suckling to the end of each movement. Simultaneous registrations have not been done with all of the muscles involved during the early stages of suckling, but one can state with certainty that the intermittent and rhythmic contractions of all of the muscles mentioned above happen in succession.

What are the consequences of these functions on the interosseous dental zone (Fig. 2.7)?

The contractions of the tongue muscles begin to predominate at the same time as the genioglossus muscles insert totally on the connective tissue of this zone of growth. This explains the sagittal orientation of the collagen fibres of the connective tissue, which is seen in the newborn only at the muscular insertions.

At the start the geniohyoid muscles insert on the two dental osseous frameworks and have an oblique posterior and outward direction, therefore involving a tendency for separation of the two dental bones at the base of the interdental space.

The platysma muscle of the chin does not insert directly under the zone of growth; on the other hand, there are insertions on to the bony edges, laterally (quantitatively weak in comparison with the lingual muscles) and corresponding to the muscular fibres oriented interiorly and laterally in the majority (the triangular muscles of the lip) and anteriorly and medially in a minority (the quadratus of the chin).

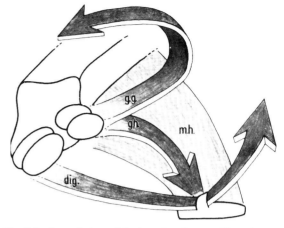

Fig. 2.7 Lateral view of the interdental space. Muscular insertion on the 'ossicles.' dig = digastric muscle; gh = geniohyoid muscle; gg = genioglossus muscle; mh = mylohyoid muscle.

During breathing and posterior lingual mandibular movement the anterior belly of the digastric contracts; however, it inserts directly into the connective tissue of the anterior pole of the interosseous dental space, at the level of the anterior and inferior 'ossicles'. The orientation of the tendinous fibres of insertion is not sagittal, as it is with the genioglossus muscles, but oblique, backward and outward.

One can see that, locally, the direct insertions of the tendinous fibres both of the genioglossus muscles in the back and the digastric muscles in front and below, are the origin of the intermittent traction and shearing forces in the IODS.

The contractions of the temporomasseter muscles and the pterygoids exert an opposing force on this median zone by means of their actions on the two ascending rami of each hemimandible:

1. lateral movement occurs by contraction of the masseter and temporal muscles;
2. medial movement is provided by the pterygoid muscles.

All these observations indicate that the connective tissue zone of the IODS is under the influence of the intermittent tensions of traction and pressure also shown by:

1. the orientation of the fibres of the connective tissue — the consequence of traction and shearing forces placed upon the structure;

2. the differentiation of the connective tissue into a secondary type of cartilaginous tissue.

The insertions, which are divergent relative to the symmetrical axis of the triangular muscles of the lip in front and below, of the geniohyoid muscles behind and below, are associated with the actions of the temperomasseteric muscles and the pterygoids, explaining the triangulation with an inferior base of the IODS.

Symphysation, which begins two to three months after birth and finishes at the end of the first year, is not an isolated phenomenon; it takes place with other changes affecting the static state of the skeleton.

1. The infant generally sits at 5 months of age, and little by little develops other movements which lead to the adoption of the standing posture towards one year of age.

2. The eruption of the deciduous dentition, which begins in a quite variable fashion around 6 months, is certainly an important factor in the stability of the symphyseal region: the establishment of the dental arches in the upper and lower incisors is the beginning of the formation of a proper incisal occlusion.

At the same time, the diet of the infant changes. Thus begins a period of transition between the infantile suction form of swallowing and the adult form of swallowing which involves chewing and a radical change in the motor strategies of mandibular movement.

Also, the metopic suture closes between 6 and 12 months of age, though the remainder of sutural synostosis of the membranous bones comes later.

CONCLUSION

1. The microscopic study of human chins in the perinatal period

This has made it possible to show that at the mesial pull of the dental bones, ossification is also endochondral, with the development of secondary cartilage appearing in the fifth month in utero and disappearing at the time of symphysation.

The structure described above is spatially more complex than the area of the mandibular condyles due to:

a. its midline location between two symmetrical dental bones;
b. the numerous muscular insertions in this region.

The *direct insertions on the zone of growth are those of the genioglossus and digastric muscles*, and also partially the geniohyoid muscles; the insertions at a distance are those of the platysma and masticatory muscles.

The tensions of traction, shearing and intermittent pressure that the muscles exert on the connective tissue zone of the IODS during suction and swallowing in the fetus and later in the newborn, are the causes of:

a. the rectilinear form of the interosseous dental space in its upper one-third, forming a triangle with an inferior base in the lower two-thirds of the space;
b. the orientation of the collagen fibres in the connective tissue;
c. the appearance of the borders of secondary cartilage or isolated nodules of cartilage;
d. enchondral ossification in the zones where hydrostatic pressure predominates in a non-intermittent fashion.

2. What is the role of the growth of the mandible associated with this enchondral ossification?

It is certainly the development of the entire mental trigone; but the secondary cartilaginous border exists along the entire upper border of the dental bones and one can attribute to it an effect on the growth of the horizontal rami of the mandible in their anterior segment. We do not have at our disposal a comparative measurement taken from a subject between 6 months in utero and 12 months postnatal.

The measurements taken from the mandibles in M. Augier's collection in the Orphila Museum (Faculty of Medicine of Paris) make it possible to state that the intercanine distance exceeds 12 mm in utero and 28 mm at 6 months postnatal.

The presence of ossicles is associated with the entire anatomico–functional structure already described and is directly related to suction and swallowing in the fetus and the newborn. These ossicles are specific to modern man. Le Double, in his treatise on the variations of the bones of the face, reported that they have not been found in primates other than man. Bourgerette (1908) examined a large number of mandibles of vertebrates including anthropomorphic monkeys; he wrote:

We can therefore affirm that the bones of the chin do not exist in other animals and that they constitute a special character of the human species.

These observations show the importance of the lingual insertions to the connective tissue of the interdental space and the genesis of the structure of the chin. It was for this reason that we proposed the term 'lingual bone' (Ginisty 1981).

The morphogenesis of the chin is of an epigenetic nature, under the influence of local and regional primordial conditions of a functional, biomechanical nature in the oral development of the fetus and the infant.

A comparison of these anatomico–functional relationships with those of other primates may, perhaps, make it possible to arrive at an understanding of the specific human structures which are called the 'ossicles' of the chin.

BIBLIOGRAPHY

Anderson L D 1965 Compression plate fixation and the effect of different types of internal fixation on fracture healing. Journal of Bone and Joint Surgery 47-A: 191–208

Anderson G C, Vidyasagar D 1979 Development of sucking in premature infants from 1 to 7 days post birth. Birth defects, XV, 7: 145–171, A. R. Liss, New York

Ashton B A, Allen T D, Howlett C R, Eaglesom C C, Hahori A, Owen M 1980 Formation of bone and cartilage by marrow stromal cells in diffusion chambers in vivo. Clinical Orthopaedics and Related Research 151: 294–307

Augier M 1931 Développement de la mandibule. In: Poirier P, Charpy A (eds) Traité d'anatomie humaine. Masson, Paris, pp 488–497

Basset C A L, Herrmann I 1961 Influence of oxygen concentration and mechanical factors on differenciation of connective tissues in vitro. Nature 190: 460–461

Beecher R M 1977 Function and fusion at the mandibular symphysis. American Journal of Physical Anthropology 47: 325–336

De Beer G 1937 The development of the vertebrate skull. Oxford University Press, Oxford

von Bertolini R, Wendler D, Hartmann E 1967 Die Entwicklung der Symphysis mentis beim Menschen. Anatomischer Anzeiger 121: 55–71

Bolender CH 1972 Etude comparative du développement mandibulaire chez le fœtus du rat et chez le fœtus humain. Thèse de Médecine, Université Louis-Pasteur, Strasbourg

Bourgerette M 1908 Les os mentonniers. Thèse de Médecine, Michalon Ed., Paris

Broom N D 1980 Simultaneous morphological and stressing studies of soft connective tissues maintained in their wet functional condition. In: Parry D A D, Creamer L K (eds) Fibrous proteins: scientific, industrial and medical aspects. Vol 2. Academic Press, London, pp 89–98,

Coquerel A 1982 Contribution à l'étude de l'oralité du fœtus et du nouveau-né humain. Mémoire A.E.A. Biol. Develop., Université René-Descartes, Paris

Couly G 1980 Biomécanique osseuse maxillo-faciale. Généralités. Encyclopedie Medico Chirurgicale (Paris), Stomatologie, 22001 D15, 4

Couly G 1983 Nouvelles conceptions du syndrome de Pierre Robin: dysneurulation du rhombencéphale. Revue de Stomatologie 84: 225–232

Darque J 1972 La tétée du nourrisson. La Clinique, LXVII, 687: 305–310

Du Brul E L, Sicher H 1954 The adaptative chin. Thomas Ed., Springfield, pp 15–18, 83–89

Flint M H, Gillard G, Merrilees M J 1980 The effects of local physical environmental factors on connective tissue organization and glycosaminoglycan synthesis. In: Parry D A D, Creamer L K (eds) Fibrous proteins: scientific, industrial and medical aspects. Vol 2. Academic Press, London, pp 107–119

Glucksmann A 1939 Studies on bone mechanics in vitro. II. The role of tension and pressure in chondrogenesis. Anatomical Record 73: 39–51

Gryboski J D 1969 Suck and swallow in the premature infant. Pediatrics 43(1): 96–102

Hall B K 1967 The formation of adventicious cartilage by membrane bones under the influence of mechanical stimulation applied in vitro. Life Sciences 6: 663–667

Hall B K 1970 Differentiation of cartilage and bone from common germinal cells. Journal of Experimental Zoology 173: 383–394

Hall B K 1971 Histogenesis and morphogenesis of bone. Clinical Orthopaedics and Related Research 74: 249–268

Hall B K 1972 Immobilization and cartilage. Transformation into bone in the embryonic chick. Anatomical Record 173: 391–404

Judet J, Judet R 1962 L'ostéogénèse et les retards de consolidation et les pseudarthroses des os longs. In: VIIIᵉ Congrès S.I.C.O.T., pp 315–458

Kernek C B, Wray J B 1973 Cellular proliferation in the formation of fracture callus in the rat tibia. Clinical Orthopaedics and Related Research 91: 197–209

Ketenjian A Y, Charalampos A 1975 Morphological and biochemical studies during differentiation and calcification of fracture callus cartilage. Clinical Orthopaedics and Related Research 107: 266–273

McKibbin B 1978 The biology of fracture healing in long bones. Journal of Bone and Joint Surgery 60-B: 150–162

Kummer B 1963 Principes de la biomécanique de l'appareil de soutien et loco-moteur de l'homme. IXᵉ Congrès

S.I.C.O.T. Cours de perfectionnement, 1963. Verlag der Wiener Medizinischer Akademie, Vienna

Le Diascorn H 1972 Anatomie et physiologie des sutures de la face. Ed. Prélat, Paris

Le Douarin N 1971 Caractéristiques ultrastructurales du noyau interphasique chez la caille et le poulet et utilisation des cellules de caille comme "marqueurs biologiques" en embryologie expérimentale. Annales d'Embryologie et de Morphogenese 4: 125–135

Le Double A F 1906 Traité des variations des os de la face de l'homme et de leur signification au point de vue de l'Anthropologie zoologique. Ed. Vigot, Paris

Le Liévre C H 1974 Rôle des cellules mésectodermiques issues des crêtes neurales céphaliques dans la formation des arcs branchiaux et du squelette viscéral. Journal of Embryology and Experimental Morphology 31(2): 453–477

Le Liévre C H 1976 Contribution des crêtes neurales à la génèse des structures céphaliques et cervicales chez les oiseaux. Thèse Doctorat Sciences, C.N.R.S., n° A.O, 12279, Paris

Maronneaud P L 1948 L'ossification des formations cartilagineuses du 1er arc branchial chez l'homme. 1re partie. Revue d'Odonto-stomatologie (Bordeaux), VI: 42–57

Millard D R, Lehman J A, Deane M, Garts W P 1971 Median cleft of the lower lips and mandible: a case report. British Journal of Plastic Surgery 24: 391–395

Millard D R, Wolfe S A, Berkowitz S 1979 Median cleft of the lower lip and mandible: correction of the mandibular defect. British Journal of Plastic Surgery 32: 345–347

Mindell E R, Rodbard S, Kwasman B G 1971 Chondrogenesis in bone repair. Clinical Orthopaedics and Related Research 79: 187–196

Monroe C W 1966 Midline cleft of the lower lip mandible and tongue with flexion contracture of the neck: case report and review of the literature. Plastic and Reconstructive Surgery 38: 312–319

Murray P D F, Drachman D B 1969 The role of movement in the development of joints and related structures: the head and neck in the chick embryo. Journal of Embryology and Experimental Morphology 22(3): 349–371

Petrovic A, Stutzmann J 1972 Le muscle ptérygoîdien externe et la croissance du condyle mandibulaire. Recherches expérimentales chez le jeune rat. Orthodontie Francaise 43: 271–285

Petrovic A, Stutzmann J 1979 a Contröle de la croissance post-natale du squelette facial. Données expérimentales et modèle cybernétique. Actualites Odontostomatologiques 128: 811–841

Petrovic A, Stutzmann J 1979 b A cybernetic view of facial growth mechanism in long term treatment in cleft lip and palate. Proceedings of the 1st International Symposium. Huber publishers, pp 14–55

Piveteau J 1963–64 La grotte du Regourdou (Dordogne). Paléontologie humaine. Ann. de Paléontologie (Vertébrés), XLIX, 285–304, VL, 159–194

Pritchard J J, Scott J H, Girgis F G 1956 The structure and development of cranial and facial sutures. Journal of Anatomy 90: 73–89

Raimbault J, Le Moing G, Laget P 1977 Etude électromyographique de la déglutition chez le jeune enfant. Annales de Pediatrie 24: 6, 7, 459–465

Raimbault J, Le Moing G, Laget P 1979 Enquête sur l'électromyographie des troubles de la déglutition chez l'enfant. Pédiatrie, XXXIV 7, 681–693

Rey A, Vasquez M P, Jennequin P, Marie M P 1982 Fentes labio-mandibulaires. A propos d'un cas. Revue de la littérature. Revue de Stomatologie et de Chirurgie Maxillo-faciale 83, 1, 39–44

Robin P 1928 La glossoptose. Doin, Paris, pp 40–43

Rouviére H 1939 Anatomie générale. Masson, Paris

Sarmiento A, Schaeffer J F, Beckerman-Latta L, Enis J E 1977 Fracture healing in rat femora as affected by functional weight-bearing. Journal of Bone and Joint Surgery 59-A, 3: 369–375

Senecail B 1979 L'os hyoïde. Introduction anatomique à l'étude de certains mécanismes de la phonation. Mémoire D.E.R.B.H., UER des St. Pères, Paris

Sherman J E, Goulian D 1980 The successful one-stage surgical management of a midline cleft of the lower lip, mandible, and tongue. Plastic and Reconstructive Surgery 66, 5: 756–759

Stutzmann J, Petrovic A 1976 Experimental analysis of general and local extrinsic mechanisms controlling upper jaw growth. In: factors affecting the growth of the midface. McNamara, Ann Arbor, Michigan, pp 205–237

Sulas V 1967 Réparation du bec-de-lièvre inférieur avec bifidité de l'arc mandibulaire. Annales de Chirurgie Infantile 8(3): 233–236

Ten Cate A R, Freeman E, Dickinson J B 1977 Sutural development and structure. American Journal of Orthodontics 71, 6: 622–636

Urist M R, Wallace Th H, Adams Th 1965 The function of fibro cartilaginous fracture callus. Journal of Bone and Joint Surgery 47-B, 2: 304–318

Weidenreich F 1936 The mandibles of sinanthropus pekinensis. A comparative study. Paleontologia Sinica, VII, D4: 31–43

3. New biological information on the morphogenesis of the mandible

A. Petrovic   J. Stutzmann

Existing knowledge on the postnatal growth of the mandible was radically transformed when it was shown, first, that the growth of the condylar cartilage in mammals, including man, can be affected by biomechanical factors (in which the condylar cartilage, a type II cartilage (Fig. 3.1) is distinguishable from all the other growing cartilages deriving from the primary skeletal cartilage of the fetus), and second, that the biological potential of growth and of response to different types of treatment of the human mandibular tissue is quite variable from one individual to another (Petrovic & Charlier 1967, Petrovic et al 1975, Petrovic et al 1985, Petrovic et al 1986). A new paradigm was thus introduced into the body of knowledge on dentofacial orthopaedics and maxillofacial surgery.

The two biological particularities mentioned above should be explained:

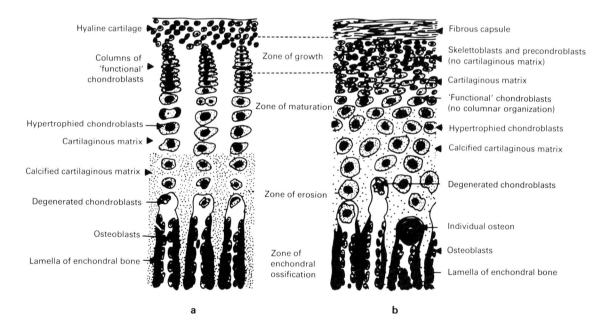

Fig. 3.1   a. Histological structure of conjugal cartilage of a long bone (type 1 cartilage). The growth compartment is made up of condrablasts, cells which divide *and* synthesize the specific cartilaginous matrix. The cartilage which surrounds the condrablasts belongs to type 2. The collagen which surrounds the hypertrophied condrablasts belongs to type X *and* type 2. **b.** Histological structure of condylar cartilage (type 2 cartilage). The growth compartment consists of skelettoblasts and precondrablasts, cells which divide but do *not* synthesize the cartilaginous matrix. The collagen which surrounds them is of type 1. The condrablasts synthesize the specific cartilaginous matrix but do not undergo further divisions. The cartilage which surrounds them is of type 2 *and* type 1. The collagen which surrounds the hypertrophied condrablasts belongs to type X *and* type 2.

17

1. it is possible to stimulate (Charlier et al 1968, Charlier et al 1969a, Petrovic et al 1975) or restrain (Charlier et al 1969b, Petrovic et al 1975) the growth of the mandible by means of functional methods or appropriate orthopaedic techniques;

2. the effect obtained varies tremendously from one individual to another, but interindividual variation, far from being a haphazard event, in fact obeys precise biological laws. The phenomenological expression of these biological laws lends itself in contemporary clinical practice to identification and delineation by cephalometric analysis (Petrovic et al 1986).

CELLULAR ASPECTS OF THE GROWTH OF THE MANDIBLE. THE SKELETOBLAST AS A CELLULAR PRECURSOR OF THE PREOSTEOBLAST, PREOSTEOCLAST AND THE TYPE II PRECHONDROBLAST

Classically, the precursor cell of bone, the osteoprogenitor cell, can differentiate either into a preosteoblast (the line of cells which form bone) or a preosteoclast (the line of cells which resorb bone).

Our research has shown that the precursor cell of bone is in reality a 'skeletoblast,' that is to say a cell capable of dividing fifty times and differentiating not only into a preosteoblast (which evolves into an osteoblast and then an osteocyte) and into a preosteoclast (giving rise, by fusion, to the osteoclast), but also into a type II prechondroblast evolving into a type II chondroblast. The preosteoblast and the type II prechondroblast only divide between five and eight times[*]. At the level of the alveolar bone, the skeletoblast is not the only source of the preosteoclast or the preosteoblast. The precursors of these two cell varieties can also come from the hematopoietic organs and arrive through the circulatory system. It has been established that every dental movement causing heavy and continuous forces exerts a deleterious effect on the vascularization of the periodontium

and in this way compromises the blood-born arrival of the precursor cells of the osteoclast and the osteoblast.

The differentiation of the skeletoblast into a preosteoblast is characterized by subperiosteal, appositional growth of the mandible, as well as simultaneous renewal of the alveolar bone. The differentiation of the skeletoblast into a type II prechondroblast is characteristic of the condylar and coronoidal cartilage (Fig. 3.1), the cartilaginous tissue which can form in the callus of a fracture (Petrovic & Stutzmann 1981, Stutzmann & Petrovic 1982), as well as in certain varieties of sarcomas of the mandible (Stutzmann et al 1980a, Petrovic & Stutzman 1980, Petrovic 1982).

Whether the skeletoblast differentiates into a preosteoblast (in the case of bone) or into a type II prechondroblast (in the case of condylar cartilage or the cartilage in the callus after a fracture), the rate and total quantity of cellular divisions depend on *local* biochemical, bioelectrical and biomechanical factors. This explains the fact that it is possible to stimulate or to restrain the growth of condylar cartilage along the posterior border of the mandible. This can be done by using an appropriate functional orthopaedic apparatus which modifies the activity of local muscles and also creates electrical and magnetic fields. In the future it will be possible to add to this list chemical substances which can amplify or attenuate the action of one or other 'mediator' or physiological 'regulator' acting in the causal chain controlling cellular division (Stutzmann 1986).

BIOLOGICAL CHARACTERISTICS OF CONDYLAR CARTILAGE

The concept that the growth of condylar cartilage can be affected by means of orthopaedic and functional methods arose from experiments done on young rats (Charlier et al 1968, Charlier et al 1969a, Charlier et al 1969b; Petrovic et al 1975) and later corroborated in the monkey (Stockli & Willert 1971, McNamara & McBride 1974, McNamara et al 1975, Komposch & Hockenjos 1977). Does this hold also for the human species? Our response is categorically positive for the following reason: at the tissue, cellular and molecular levels, the response of different growing cartilages

---

[*]It is for this reason that a cranial suture, in which (most often for a genetic cause) the skeletoblasts *all* differentiate at an early stage of ontogenesis into preosteoblasts, undergoes premature fusion.

obtained from the infant to biomechanical factors does not differ significantly from that of similar growing cartilage taken from the animal (Petrovic 1982, Stutzmann & Petrovic 1982, Petrovic 1984a, b, 1985, Petrovic & Stutzmann 1984). In effect, light pressure exerted, in an organ-typical culture, on a growing condylar cartilage or on cartilage from the callus of a fracture, coming either from different laboratory animals (the rat, the mouse, the guinea pig, the rabbit, and the *Saimuru sciureus* monkey) or the infant, causes the following systematic variations in the cells of the mitotic compartment (Table 3.1):

1. the activity of the ATPases associated with cellular membranes is modified: ATPase activity associated with [Na$^+$, K$^+$] increases, and the ATPase activity associated with [Ca^{++}, Mg^{++}] and [H$^+$] diminishes;
2. the cellular [Na$^+$] diminishes, and the cellular [Ca^{++}] and [H$^+$] increase. The intracellular pH decreases;
3. the number of cellular divisions diminishes.

These variations have not been seen when cartilage of the primary type (the epiphyseal cartilage of the long bones, the cartilage of the metatarsals and metacarpals, and the spheno-occipital synchondrosis) taken from the same mammals, including the human infant, has been exposed in organ-typical culture to similar or even greater pressures (Petrovic 1982, 1984a, 1984b, 1985).

In other words, the ability to respond to a biomechanical factor is tissue specific (the response is possible when the action is on condylar cartilage but it does not occur with so called primary cartilage) and does not depend on the species of animal. One cannot say that the condylar cartilage of an infant should respond to a functional or orthopaedic apparatus in a *different* manner to the condylar cartilage of mammals such as the rat or the monkey.

The cells in a very narrow zone of the osteochondral junction of the human rib, of the rat or of the dog, placed in organ-typical tissue culture, react to pressure by the same cellular and molecular variations as the cells of the mytotic

Table 3.1 Schematic representation of the effects of continued pressure on the cells of the mitotic compartment of cartilages placed on organ-typical culture and arising either from the infant, or from different laboratory animals (rat, mouse, guinea pig, rabbit, *saimuru sciureus* monkey). Only the cartilage of fractured callus and the zone of differentiation of precursor chondroblasts into progenitor chondroblasts shows a biological behaviour similar to that of the cells of the mitotic compartment of chondylar cartilage.

	Chondylar cartilage of the mandible	Cartilage of the callus of a fracture	Zone of precursor chondroblasts	Osteochondral junction of rib Zone of differentiation of precursor chondroblasts and progenitor chondroblasts	Zone of primary chondroblasts deriving from progenitor chondroblasts	Growth cartilages — long bones — metatarsals — metacarpals — spheno-occipital synchondrosis
ATPase activities associated with the membrane						
ATPase [Na$^+$/K$^+$]	↗	↗	--→	↗	--→	--→
ATPase [Ca^{++}, Mg^{++}]	↘	↘	--→	↘	--→	--→
ATPase [H$^+$]	↘	↘	--→	↘	--→	--→
Cellular concentration						
Na$^+$	↘	↘	--→	↘	--→	--→
Ca^{++}	↗	↗	--→	↗	--→	--→
H$^+$	↗	↗	--→	↗	--→	--→
Intracellular pH	↘	↘	--→	↘	--→	--→
Cellular volume	↘	↘	--→	↘	--→	--→
Intracellular H$_2$O	↘	↘	--→	↘	--→	--→
Intracellular AMPc	↗	↗	--→	↗	--→	--→
Number of cellular divisions	↘	↘	--→	↘	--→	--→

→ important variation, → median variation, --→ insignificant variation

compartment of the condylar cartilage (Petrovic et al 1983, Petrovic 1984a, b, Graber et al 1985). In the situation where the osteochondral junction of the human rib can serve as a substitute tissue for condylar cartilage, this observation is of definite interest in maxillofacial surgery. The growth cartilage at the sternal end of the clavicle also shows a biological behaviour similar to that of condylar cartilage.

The application of a functional apparatus causes, both in the young rat and in the human infant, an increase in cellular Na^+ and intracytoplasmic water, firstly in the cells of the mytotic compartment of the condylar cartilage and secondly in the cells of the subperiosteal zone along the posterior border of the ascending ramus. A similar phenomenon can be demonstrated by nuclear magnetic resonance which has become a new tool in the estimation and prediction of therapeutic effectiveness.

GROWTH OF THE HUMAN MANDIBLE: THE BIOLOGICAL BASIS OF THE VARIABILITY BETWEEN INDIVIDUALS

Our studies on the rate of growth of the mandible evaluated by the measure of the mitotic index at the subperiosteal level along the lateral surface of the ascending ramus, have led to the following observation (Petrovic & Stutzmann 1982, Petrovic et al 1985): the mitotic index is considerably more elevated on a mandible with an anterior growth rotation than on a mandible with a posterior growth rotation. However, this observation only holds statistically; many exceptions are seen in individual cases (Stutzmann et al 1979, 1980b, Stutzmann & Petrovic 1980, 1981, 1983, 1984).

In another study, we listed the results relative to the rate of regeneration of human alveolar bone in six categories (Petrovic et al 1985). The same categorization was then applied with success in our research on the mytotic index of the superiosteal layer (Petrovic et al 1986). It emerged from these two studies that the dominant factor, not only for the growth of the mandible but also for the response to orthopaedic or orthodontic treatment, is the rate of subperiosteal growth of the bone and the speed of regeneration of the alveolar bone. Thus, both the second category of growth and the

sixth category of growth can give rise — depending on the adjustment of the self-regulating system — either to a posterior rotation or to an anterior rotation. The fifth category of growth can give rise to a posterior, 'neutral' or anterior rotation of growth (Petrovic et al 1985, Petrovic et al 1986). By use of cephalometrics, Lavergne & Gasson (1982) defined 11 types of growth rotation of the face, designating each by one of three symbols:

1. A letter (P, R or A) indicating the variety of rotation (posterior, neutral or anterior);
2. A number (2, 1 or 3) representing, by analogy with the Angle classification, an estimation of the difference between the growth potentials of the maxilla and mandible (see Petrovic et al 1986);
3. A letter (D, N or M) indicating the sagittal basal relationships.

It was notable that one could establish an excellent correspondence between the 6 categories of growth which were biologically defined and the 11 types of growth rotation which were cephalometrically defined (Petrovic et al 1985, Lavergne & Petrovic 1985, Petrovic et al 1986, 1987).

The comparison between the biological and the cephalometric data also makes it possible to better understand the reason why the simple classification of growth rotations of the mandible into posterior, 'neutral' and anterior, does not take account of anthropological and clinical complexity. For example, in the category 2 of growth, the posterior rotation of the mandible can give individuals having a rotational type PlN a normal appearance or even an Angle Class 3 appearance, because the length of the mandible coming into occlusion corresponds relatively well to that of the maxilla. In reality, the growth category 2 is categorized by a relatively weak response to treatment aimed at stimulating the growth of the mandible (see Petrovic et al 1985, Lavergne & Petrovic 1985, Petrovic et al 1986).

As a result of these investigations, we have studied the index of cellular division (by marking with tritiated thymidine) in the subperiosteal layer of the mandible in three different areas: the posterolateral zone of the ascending ramus, the vestibular zone of the horizontal ramus and the area of prominence of the chin. Bone biopsies were

Table 3.2 Index of cellular divisions in the subperiosteal layer. Classification based on the category of growth and on the type of rotation of the mandible.

Young males aged 10 to 13 years

Biological category of growth	Cephalometric type of mandibular rotation	Posterolateral zone of the ascending ramus			The lateral zone of the horizontal ramus			Region of prominence of the chin		
		N	Median %	(Extreme values) %	N	Median %	(Extreme values) %	N	Median %	(Extreme values) %
1	P2D	17	2.9	(1.7–3.5)	9	0.09	(0.03–0.18)	13	0.05	(0.01–0.08)
2	A2D	11	4.3	(3.8–5.1)	5	0.11	(0.09–0.23)	11	0.08	(0.02–0.13)
2	P1N	9	4.6	(4.1–4.9)	5	0.14	(0.08–0.27)	5	0.09	(0.05–0.18)
3	R2D	13	5.2	(4.6–5.9)	7	0.12	(0.09–0.25)	5	0.11	(0.07–0.20)
4	R1N	11	6.3	(5.8–7.0)	7	0.14	(0.07–0.26)	3	0.19	(0.10–0.25)
5	A1D	19	7.5	(6.9–8.3)	11	0.15	(0.11–0.28)	7	0.16	(0.12–0.27)
5	A1N	13	7.8	(7.3–8.6)	7	0.18	(0.13–0.27)	5	0.19	(0.13–0.30)
5	P1M	9	7.9	(7.1–8.2)	5	0.19	(0.16–0.32)	7	0.20	(0.12–0.31)
5	R3M	9	8.0	(7.2–8.9)	7	0.21	(0.15–0.29)	9	0.24	(0.15–0.33)
6	A3M	3	9.8	(9.4–10.9)	3	0.37	(0.32–0.48)	11	0.47	(0.42–0.59)
6	P3M	5	11.1	(9.7–12.3)	3	0.32	(0.30–0.43)	9	0.49	(0.41–0.66)

carried out at the time of surgical operations on young males between 10 and 13 or between 17 and 20 years of age. The cases were classified depending on the category of growth and the type of rotation. The results are shown in tables 3.2 and 3.3, which the reader should examine in detail to extract all the information. We will not go into this specifically except to point out several significant results.

1. When dealing with young males between the age of 10 and 13 or between 17 and 20, there exists (when one examines the 'categories of growth' from 1 to 6) a remarkable correspondence between the variations of the rate of subperiosteal ossification, a. at the level of the posterolateral zone of the ascending ramus, b. at the level of the vestibular zone of the horizontal ramus and c. at the level of the zone of prominence of the chin. We should recall the parallel which has already been drawn between the variations in the rate of subperiosteal ossification at the level of the posterolateral zone of the ascending ramus and the variations in the rate of regeneration of the alveolar bone.

2. Whatever the category of growth being considered, the rate of subperiosteal ossification is

Table 3.3 Index of cellular divisions in the subperiosteal layer. Classification based on the category of growth and on the type of rotation of the mandible.

Young males aged 17 to 20 years

Biological category of growth	Cephalometric type of mandibular rotation	Posterolateral zone of the ascending ramus			The lateral zone of the horizontal ramus			Region of prominence of the chin		
		N	Median %	(Extreme values) %	N	Median %	(Extreme values) %	N	Median %	(Extreme values) %
1	P2D	13	0.13	(0.02–0.26)	7	0.002	(0.000–0.005)	9	0.011	(0.005–0.023)
2	A2D	9	0.28	(0.13–0.39)	5	0.005	(0.003–0.008)	9	0.026	(0.016–0.034)
2	P1N	9	0.32	(0.24–0.47)	5	0.006	(0.004–0.011)	7	0.022	(0.016–0.055)
3	R2D	11	0.40	(0.29–0.56)	9	0.006	(0.004–0.010)	5	0.059	(0.047–0.086)
4	R1N	7	0.67	(0.53–0.88)	7	0.010	(0.008–0.013)	7	0.078	(0.065–0.101)
5	A1D	9	0.88	(0.66–1.15)	9	0.011	(0.008–0.014)	5	0.123	(0.118–0.149)
5	A1N	11	1.05	(0.79–1.23)	7	0.014	(0.009–0.018)	5	0.136	(0.125–0.168)
5	P1M	7	1.69	(0.80–1.92)	5	0.023	(0.010–0.026)	9	0.197	(0.134–0.279)
5	R3M	7	1.31	(0.98–0.57)	9	0.015	(0.008–0.019)	13	0.161	(0.142–0.283)
6	A3M	3	1.64	(1.45–1.76)	7	0.019	(0.015–0.026)	15	0.372	(0.257–0.686)
6	P3M	3	1.78	(1.51–1.82)	9	0.021	(0.013–0.028)	11	0.398	(0.281–0.576)

always more elevated in the posterolateral zone of the ascending ramus than in the vestibular zone of the horizontal ramus. This can be explained by the fact that in the vestibular zone bony apposition takes place exclusively as a thickening of the mandible, whereas in the posterolateral zone of the ascending ramus there is both thickening and posterior growth. The patterns of these observations are particularly clear in the 10 to 13 year old age group, but also remain valid in the 17 to 20 year age group.

3. In the three regions of the mandible, the rate of subperiosteal ossification is considerably less elevated in the 17 to 20 year age group than in the 10 to 13 year age group. It is less elevated, but it is still relatively significant, particularly when one takes into consideration the categories of superior growth (categories 5 and 6). In the rotational type P1M, and to a lesser degree in the rotational type R3M, the rate of subperiosteal apposition in the 17 to 20 year age group remains more elevated than in the two other rotational types (A1D and A1N) in category 5. This fact is of interest to the clinician — both the orthodontist and the maxillofacial surgeon — because from age 10 to 13 onwards, in diagnosing a rotational type P1M and R3M, one should foresee the eventuality of a *prolonged* growth of the mandible.

4. Between the age of 10 and 13, the rate of subperiosteal ossification of the chin is (in comparison with the vestibular zone of the horizontal ramus of the mandible) *weaker* in the growth categories 1 and 2 and more *elevated* in the growth category 6. In categories 3, 4 and 5 no significant difference is apparent between these two regions of the mandible.

In the 17 to 20 year age group, the bony apposition in all categories of growth is clearly more elevated in the region of the chin than in the vestibular region of the horizontal ramus.

5. The analysis of tables 3.2 and 3.3 shows that subperiosteal apposition is only a contributing factor to the formation of the chin, this contribution becoming greater and greater as one passes from growth category 1 to growth category 6. Another phenomenon contributes to the formation of the chin: in a large percentage of cases of anterior growth rotation we have observed, in analyzing the process of bone formation and resorption distal

and mesial to extracted premolars, the existence of distalization of the inferior dental arch (Stutzmann et al 1979, 1980b, Stutzmann & Petrovic 1980, Stutzmann & Petrovic 1984). It should be emphasized that, statistically, the existence of this distalization in the case of anterior growth rotation of the mandible is more frequent and its intensity more pronounced the greater the category of growth (see Graber et al 1985).

THE THERAPEUTIC EFFECTIVENESS OF ORTHODONTIC, ORTHOPAEDIC OR FUNCTIONAL APPLIANCES AS A FUNCTION OF THE RATE OF SUBPERIOSTEAL OSSIFICATION OF THE ASCENDING RAMUS OF THE MANDIBLE (POSTEROLATERAL ZONE)

It emerged from our research, both in the rat and in man, that in the mandible of any particular individual, there exists a parallel between the level of the rate of subperiosteal ossification, the level of the growth rate of the condylar cartilage and the level of the rate of regeneration of alveolar bone (Petrovic et al 1986).

In studying the rate of regeneration of alveolar bone in the mandible from one infant to another, we noted that passing from growth category 1 to growth category 6, it becomes easier to stimulate the sagittal growth of the mandible by means of functional methods, and increasingly difficult to retard it by means of orthopaedic appliances. Recourse to maxillofacial surgery is very frequent in category 1 and almost indispensable in category 6 (Petrovic et al 1986).

What happens when the category of growth is evaluated not just by the rate of regeneration of alveolar bone, but also by the rate of apposition of subperiosteal bone?

Bone biopsies were carried out during surgical interventions for fracture in males between the ages of 10 and 13. The bony fragment was placed in an organ-typical tissue culture. After one hour of culture in the presence of tritiated thymidine (a specific precursor of deoxyribonucleic acid), the osseous fragment was fixed and mounted. The histological sections were audioradiographed in order to show the cells which were synthesizing deoxyribonucleic acid. In this manner the index of cellular divisions can be evaluated. The type of rotation was defined cephalometrically using the classification of Lavergne & Petrovic (1985). Clinical effectiveness after one year of treatment was estimated according to the methods described by Petrovic et al (1986). Only those cases where the fracture took place 1 to 2 years before the begin-

ning of the orthodontic or orthopaedic treatment were retained.

The results of this study are shown in table 3.4. Examination of the table shows evidence of the clinical efficacy of the treatment, evaluated one year after the onset of treatment, and that this depended much more on the biological category of growth than on the variety of treatment. In effect,

Table 3.4 Therapeutic effectiveness of an orthodontic, orthopaedic or functional apparatus as a function of the extent of subperiosteal ossification of the ascending ramus of the mandible (posterolateral zone).

Biological category of growth	Type of rotation	Index of cellular division	Type of treatment	Clinical effectiveness after one year of treatment
1	P2D	1.9	Edgewise	+
		3.0	LSU Activator	+
		2.8	Bionator	+
		2.1	Bionator	+
		2.3	Fraenkel	0
2	A2D	3.8	Edgewise	+
		4.9	LSU Activator	+
		4.6	Bionator	+(+)
		4.2	Fraenkel	+
		4.1	Herbst	+ +
2	P1N	4.6	Edgewise	+
		4.7	Active retropulsion of the mandible	+ +
		4.8	Fraenkel	+ +
		4.1	Fraenkel	+ +
3	R2D	5.0	Edgewise	+(+)
		5.3	LSU Activator	+(+)
		4.8	Fraenkel	+ +
		5.6	Begg	+ +
4	R1N	6.6	LSU Activator	+ +
		6.3	Bionator	+ +
		5.9	Fraenkel	+ +
5	AID	7.9	LSU Activator	+ + +
		7.5	Bionator	+ + +
		6.9	Bionator	+ + +
		7.2	Begg	+ + +
		8.1	Bimler	+ + +
		7.6	Herbst	+ + +
5	AIN	7.6	Edgewise	+ + +
		8.5	LSU Activator	+ + +
		7.7	Bionator	+ + +
		8.0	Begg	+ + +
		8.2	Bimler	+ + +
5	PIM	7.8	Active retropulsion of the mandible	+ +
		7.2	Fraenkel	+ +
5	R3M	7.9	Fraenkel	+ +
		8.7	Begg	+(+)
6	P3M	10.6	Fraenkel	+
		11.2	Active retropulsion of the mandible	+
6	A3M	10.9	Fraenkel	0

The therapeutic effectiveness was evaluated semiquantitatively, in 39 young male patients aged 10 to 13 years, by the comparison of the observed result with the expected result. The following scale was adopted: 0 = without effect; + = very limited effect; + + = suboptimal effect (the treatment should be continued); + + + = optimal effect.

in passing progressively from category 1 to category 6, it became increasingly easy to correct a mandibular retrognathia and more and more difficult to correct a mandibular prognathism whether the treatment was orthodontic, orthopaedic or functional. In several cases, surgical treatment of a fracture was the occasion to correct a retrognathia in young male patients belonging to growth category 1 and to correct a prognathia associated with growth category 6.

All the results presented in this chapter corroborating our research show the existence of a correlation between the different parameters evaluating the physiological potential of growth as a response of the mandibular tissues. Certainly, the methods used in this investigation were not part of everyday clinical practice. Whilst awaiting the new procedures of medical imaging which will permit quantitative evaluation of the processes of growth and regeneration at the mandibular level in a noninvasive manner, one can meanwhile now very satisfactorily estimate individually, the biological category of growth by using cephalometrics, and more specifically, the cephalometric analysis described by Lavergne & Gasson (1982). Correspondence between the biological data and the cephalometric data is described in detail in the work of Petrovic et al (1986).

FINAL COMMENTS

Biological analysis has upset some of our understanding of the morphogenesis of the human mandible. It has made it possible to replace supposition and dogma with well-grounded medical concepts. The process of medical decision making, based on the criteria of clinical efficacy, has been enriched by recourse to these scientific criteria. In the bubbling up of ideas and working hypotheses which give rise to the richness of medical and surgical thought, biological analysis has played, according to Petrovic (1983), the role that the philosopher Karl R. Popper attributed to scientific research: the separation away of erroneous theses so as to preserve only those theories and concepts which have resisted, even if *only for the time being*, efforts at refutation. The originality of the results of the research presented in this short chapter rests foremostly in the analysis of the physiological and physiopathological phenomena which occur in the malrelations between the jaws and the malocclusions *at different levels of biological organization* (Petrovic 1982, 1984a, 1985).

The practical importance of this research is to serve as a conceptual tool for the clinician, whether it helps an individual patient or resolves a general problem in maxillofacial surgery or dentofacial orthopaedics (Petrovic et al 1986).

Biological analysis shows that subperiosteal ossification always contributes to the formation of the chin, particularly in the postpubertal period (17 to 20 years of age). But in the majority of cases, the mental protuberance results in large part from the regression of the inferior dental arch which has been shown to be the case by the study of the process of formation and resorption of bone on the mesial and distal sides of the premolars.

BIBLIOGRAPHY

Charlier J P, Petrovic A, Herrmann J 1968 Déterminisme de la croissance mandibulaire: effets de l'hyperpropulsion et de l'hormone somatotrope sur la croissance condylienne de jeunes rats. Orthodontie Francaise 39: 567–579

Charlier J P, Petrovic A, Herrmann-Stutzmann J 1969a Effects of mandibular hyperpropulsion on the prechondroblastic zone of young rat condyle. American Journal of Orthodontics 55: 71–74

Charlier J P, Petrovic A, Linck G 1969b La fronde mentonnière et son action sur la croissance mandibulaire. Recherches expérimentales chez le rat. Orthodontie Francaise 40: 99–113

Graber T M, Rakosi T, Petrovic A 1985 Dentofaccial orthopedics with functional appliances. Mosby, St-Louis (USA) p 496

Komposch D, Hockenjos C 1977 Die Reaktionsfähigkeit des temporomandibularen Knorpels. Fortschritte der kieferorthopadie 38: 121–132

Lavergne J, Gasson N 1982 Analysis and classification of the rotational growth pattern without implants. British Journal of Orthodontics 9: 51–56

Lavergne J, Petrovic A 1985 Pathogenesis and treatment conceptualization of dentofacial malrelations as related to the pattern of occlusal relationship. In: Dixon A D, Sarnat B G (eds) Second International Conference on Clinical Factors and Mechanisms Influencing Bone Growth. A Liss, New York pp 393–402

McNamara J A Jr, MrBride M C 1974 A histological study of the postnatal development of the temporomandibular articulation in Macaca mulatta. Anatomical Record 178: 408

McNamara J A Jr, Connelly T G, McBride M C 1975 Histological studies of temporomandibular joint adaptations. In: McNamara J A, Jr (ed) Control Mechanisms in Craniofacial Growth. Monograph 3, Craniofacial Growth Series, Center for Human Growth and Development, The University of Michigan, Ann Arbor, Michigan pp 209–227

Petrovic A 1982 Postnatal growth of bone: a perspective of current trends, new approaches, and innovations. In: Dixon A D, Sarnat B G (eds) Factors and Mechanisms Influencing Bone Growth. Progress in Clinical and Biological Research, vol. 101. A Liss, New York pp 297–331

Petrovic A 1983 Types d'explication dans les Sciences biomédicales et en Médecine. In: Séminaire sur les Fondements des Sciences; L'explication dans les Sciences de la Vie (ouvrage. collectif publié sous la direction de Hervé Barreau), Editions du CNRS, Paris. pp. 199–258

Petrovic A 1984a Zweckmässigkeit, Bedeutung und Gültigkeit experimentelle Forschung auf dem Gebiet der Kieferorthopädic und Orthodontie. Fortschritte der Kieferorthopadie 45: 165–186

Petrovic A 1984b An experimental and cybernetic approach to the mechanism of action of functional appliances on the mandibular growth. In: Malocclusion and the Periodontium. Monograph 15, Craniofacial Growth Series, Center for Human Growth and Development, University of Michigan, Ann Arbor, Michigan. pp 213–268

Petrovic A 1985 Point de vue d'un chercheur sur le rat comme modèle expérimental en orthodontie. Revue d'Orthopedie Dentofaciale 19: 101–113

Petrovic A, Charlier J P 1967 La synchondrose sphéno-occipitale de jeune rat en culture d'organes: mise en évidence d'un potentiel de croissance indépendant. Comptes Rendus, Academie des Sciences série D, 265: 1511–1513

Petrovic A, Stutzmann J 1980 Skeletoblast derived bone tissue tumors. A new classification of bone tumors. In: Donath A, Courvoisier B (eds) Third Symposium CEMO, "Bone and Tumors". Editions Médecine et Hygiène, Geneva, Switzerland

Petrovic A, Stutzmann J 1981 A cybernetic niew of facial growth mechanisms. In: Kehrer B, Slongo T, Graf B, Bettex M (eds) Long Term Treatment in Cleft Lip and Palate. H. Huber, Bern, Stuttgard, Vienna, pp 15–56

Petrovic A, Stutzmann J 1982 Teoria cibernetica del crecimiento craneofacial post-natal y mecanismos de accion de los aparatos ortopedicos y ortodonticos. Revista Association Argentina de Ortopedia Functional de los Maxilares 15: 7–93

Petrovic A, Stutzmann J 1984 Potencial de crecimiento del nivel tisular mandibular, rotacion de crecimiento y respuesta a aparatos funcionales. Orthodoncia (48) 96: 26–34

Petrovic A, Stutzmann J, Oudet C 1975 Control processes in postnatal growth of condylar cartilage of the mandible. In: McNamara J A jr (ed) Determinants of Mandibular Form and Growth. Monograph 4, Craniofacial Growth Series, Center for Human Growth and Development, University of Michigan, Ann Arbor, Michigan pp 14–57

Petrovic A, Stutzmann J, Champy M 1983 Autogreffe ostéochondrale de côte. Etude expérimentale. XXVIIIe Congrès Français et Ier Congrès Européen de Stomatologie et al Chirurgie maxillo-faciale, Paris, 21–24 septembre

Petrovic A, Stutzmann J, Lavergne J 1985 Effect of functional appliances on the mandibular condylar cartilage. In: Graber T M (ed) Physiologic Principles of Functional Appliances. Mosby, St-Louis, pp 38–52

Petrovic A, Lavergne J, Stutzmann J 1986 Tissue-level growth and responsiveness potential, growth rotation, and treatment decision. In: Sciences and Clinical Judgement in Orthodontics. Monograph 19, Craniofacial Growth Series, Center of Human Growth and Development, University of Michigan, Ann Arbor, Michigan, pp 181–223

Stockli P W, Willert H G 1971 Tissue reactions in the temporomandibular joint resulting from anterior displacement of the mandible in the monkey. American Journal of Orthodontics 60: 142–155

Stutzmann J 1986 Variations interindividuelles de la vitesse de renouvellement osseux de la mandibule. Intérêt en orthodontie du jeune adulte. Orthodontie Francaise 57(2) (in press)

Stutzmann J, Petrovic A 1980 La vitesse de renouvellement de l'os alvéolaire chez l'adulte avant et pendant le traitement orthodontique. Revue d'Orthopedie Dentofaciale 14 437–456

Stutzmann J, Petrovic A 1981 Die Umbaugeschwindigkeit des Alveolarknochens beim Erwachsenen vor und nach orthodontischer Behandlung. Fortschritte der Kieferorthopadie 42: 386–404

Stutzmann J, Petrovic A 1982 Bone cell histogenesis: the skeletoblast as a stem-cell for preosteoblasts and for secondary-type prechondroblasts. In: Dixon A D, Sarnat B G (eds) Mechanisms Influencing Bone Growth. Progress in Clinical and Biological Research, vol. 101, A Liss, New York, pp 29–43

Stutzmann J, Petrovic A 1983 Variaciones inducidas por gomas Class II en la reorganizacion del hueso alveolar de la mandibula humana. Revista Associacion Argentina de Ortopedia Functional de los Maxilares 16: 7–28

Stutzmann J, Petrovic A 1984 Human alveolar bone turn-over rate. A quantitative study of spontaneous and therapeutically-induced variations. In: Malocclusion and the Periodontium. Monograph 15, Craniofacial Growth Series, Center for Human Growth and Development, University of Michigan, Ann Arbor, Michigan pp 185–212

Stutzmann J, Petrovic A, Shaye R 1979 Analyse en culture organotypique de la vitesse de formation-résorption de l'os alvéolaire humain prélevé avant et pendant un traitement comprenant le déplacement des dents: nouvelle voie d'approche en recherche orthodontique. Orthodontie Francaise 50: 399–419

Stutzmann J, Petrovic A, George D 1980a Life cycle length, number of cell generations, mitotic index and modal chromosome number as estimated in tissue culture of normal and sarcomatous bone cells. In: Donath A, Courvoisier B (eds) Third Symposium CEMO, 'Bone and Tumors'. Editions Médecine et Hygiène, Geneva, Switzerland pp 188–194

Stutzmann J, Petrovic A, Shaye R 1980b Analyse der Resorptionsbildungsgeschwindigkeit des menschlichen Alveolarknochens, in organotypischer Kultur, entnommen vor und waehrend der Durchführung einer Zahnbewegung. Ein neuer Anblick in der orthodontischer Forschung. Fortschritte der Kieferorthopadie 41: 236–250

4. The chin and its anatomical significance. The importance of postural phenomena

J. Lévignac   J.-C. Chalaye   J.-L. Heim

We should note at the beginning that the four-legged mammal does not have a chin. The chin only appears with the biped gait and upright posture, and has become one of the characteristics of man. This shows at the outset the importance of postural phenomena, and we will now attempt to understand the how and the why, at the same time keeping to the essential points. The following areas will be examined in turn:

- The straightening of the body and the chain of vertebral curves.
- The functional adaptation characteristic of the masticatory region and associated with postural change.
- The regression of the dental arch on its osseous base from the *Australopithecus* until *Homo sapiens*.
- Cervicocephalic equilibrium.
- The appearance of anomalies in response to certain postural disequilibria.

THE STRAIGHTENING OF THE BODY, THE CHAIN OF VERTEBRAL CURVES AND THE MAINTENANCE OF HEAD POSITION

The totality of the skeletal readjustments associated with the straightening of the body and the passage to the biped gait should be noted here. Observing the transformations in the primate chain to arrive at man, one can note from ground level up that:

1. *The foot* is formed with a doubly concave planter arch. This arch itself appears to be suspended from the strap formed by the tendons of the anterior leg muscles and the long lateral perineal whose insertions extend almost all the way to the inferior surface of the tarsus (du Brul 1977). The calcaneus developed towards the rear modifies the lever effect of the Achilles tendon. The tibiotarsal joint is displaced forward so that the axis of gravity of the body on its supporting base is better situated.

2. *The pelvis* has a concave surface in front which maintains the support of the viscera. Posteriorly, it has on each side an external iliac fossa — a large surface for the insertion of the muscles of the buttocks which have become quite strong.

In profile, and particularly throughout the process of adaptation to the biped gait, one can see that the lumbosacral joint moves closer to the coxofemoral joint, being situated at some distance above and behind the femur. The vertebral column is also better placed with respect to the axis of the femur, man now standing upright.

It should be noted that the position of the pelvis depends on the planter support — with the shape of the foot and its arch in the case of flat feet, one can note an anterior version of the pelvis in which the lumbar lordosis is accentuated.

3. *The vertebral column* needs to be examined in its totality. It is composed of 33 vertebrae which articulate with one another, and resembles a 'long, resistant, flexible, bony shaft situated in the posteromedian portion of the body extending from the head which it supports down to the pelvis which supports it' (Rouvière 1943) (Fig. 4.1).

The vertebral bodies become thicker the more inferior their location as they extend down to the lumbosacral joint.

Seen in profile along its entire height, the vertebral column has four curves:

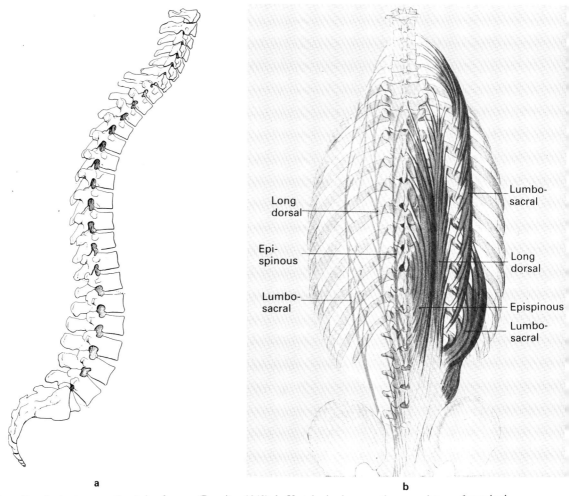

Fig. 4.1 a. Vertebral column — the chain of curves (Rouvière 1943). **b.** Vertebral column — the musculature of vertebral straightening (Rouvière 1943).

a. The dorsal curve, which is the 'primitive' curve seen in the quadruped mammal. We should note that the newborn human infant has only one dorsal curve anteriorly. When the infant begins to sit, the cervical curve forms under the action of the muscles which support the head. When the infant begins to walk, the lumbar curve appears and a chain of curves which are characteristic of man develops. In upright man, the greater dorsal curve with an anterior concavity is caused by the weight of the viscera of the thoracic cavity and the upper abdomen. It is accentuated by fatigue and by the ageing process.

b. Just below, a lumbar curve with an anterior convexity marks the straightening of the body. In fact, this convexity is in part due to the specific action of the intestinal contents and the mesenteric attachments on the lower lumbar vertebrae; the effect of straightening itself is associated with the action of the posterior spinal muscles which insert along the entire extent of the vertebral column and which also have strong inferior attachments to the iliac tuberosity and the adjacent crest. The anterior and lateral abdominal musculature provides an important counter pressure (Keith 1923).

c. The sacrum is made out of five fused vertebrae. It is attached to the iliac bones by an interlocking joint on each side, but dissipates in the curve of the anterior concavity of the pelvis. We should note that the lumbosacral joint is a 'condylar' type of joint (Dieulafé, Rouvière 1943) which makes certain tilting movements of the pelvis possible. Held in place by a substantial ligamentous structure, it transmits the weight of the trunk to the coccyx and then into the lower extremities.

d. At the top end of the vertebral column, the cervical curve with an anterior concavity places the head in the best position of equilibrium.

We should note that the centre of gravity of the head falls in front of the occipito-atlantoid axis. Added to this is an inferiorly directed tension being placed on the chin and the mandible because they support the stylohyoid arch, which partly suspends the tracheobronchial tree.* The muscles of the neck remain powerful in man, depending on the degree of anterior descent required and taking into account the dorsal curve, and tilt the head backwards to keep it straight. Thus one sees more or less marked degrees of convexity of the cervical spine to hold the head in equilibrium and to fulfill the functional requirements of the visual axis.

In reality, cervicocephalic equilibrium, as it has evolved in man, has not come to be without certain important craniofacial modifications observed during the course of primate evolution. These include, in particular, the regression of the facial prominence with the anterior movement of the oc-

cipital foramen which ameliorates the distribution of the weight of the skull on the vertebral axis.

Higher up, one can follow the vertebral column into the skull since the basi-occiput and the baso-sphenoid are cephalized vertebrae. However, we will halt at the occipital-odontoid articulation and the axis and atlas — there will be more concerning this later. Here we would like, in particular, to examine the vertebral column in its entirety as a flexible shaft, with a resilience provided by its four curves, which acts as a shock absorber for the impact of walking and which holds the head in equilibrium.

THE FUNCTIONAL ADAPTATION PARTICULAR TO THE MASTICATORY REGION AND ASSOCIATED WITH POSTURAL CHANGE

Returning to the origins of man and to postural change (tilting of the face beneath the skull), one can see that accompanying it there is a compression of the organs against the cervical spine. This creates a problem in opening the mouth if a very particular type of adaptation does not take place at the temporomandibular joint level. The mandible undergoes a luxation of a sort to open with an axis of rotation which is displaced inferiorly. Then the action of the lateral pterygoids becomes preponderant in mastication, with the alternative movements of circumduction. The composition of the forces exerted on the condyles on each side and on the two sides together tends in a way to attempt to bend the mandible in the mid portion of its arch. Here an osseous reinforcement takes place which is also a response to shearing forces (du Brul & Sicher 1954) (Fig. 4.2).

THE REGRESSION OF THE DENTAL ARCH ON ITS OSSEOUS BASE FROM AUSTRALOPITHECUS TO HOMO SAPIENS

The regression of the dental apparatus on the osseous base, a regression observed throughout the evolution of man from *Australopithecus* to *Homo sapiens*, seems to be particularly associated with changes in diet and easing of masticatory demands. With the regression of the dental arcade, the osseous base projects in profile along the entire

---

*When in the course of evolution the snout became displaced below the skull to become the face there was a compression of the organs of the neck against the vertebral column and the larynx fell as well as the hyoid creating the need for respiratory adaptation. Swallowing became easier with the new vertical position, but we should note also that with this verticality, the constrictor muscles of the pharynx, the stylopharyngeal muscle and the digastric (posterior belly) became more suspenders of the larynx and acted to elevate it. Here this action was equilibrated with that of the muscles attaching the anterior portion of the hyoid bone to the body of the mandible. The hyoid bone and the larynx are, therefore, well suspended posteriorly, as well as anteriorly.

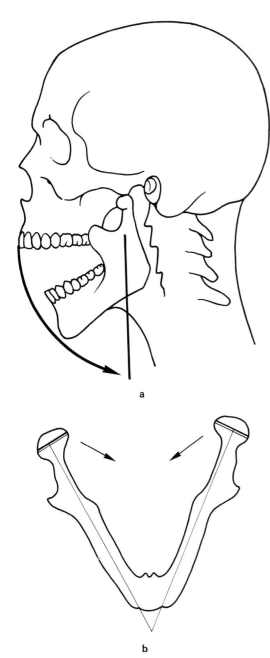

a

b

Fig. 4.2 a. During evolution, with the verticalization of the face and because of the compression of the organs against the vertebral column, one notices a transformation and 'adaptation' of the temporomandibular joint that is so effective that now the mandible undergoes a luxation of sorts to open. The function of the lateral pterygoids becomes predominant. **b.** In the centre of the mandibular arch a bony reinforcement is created which responds to the vectors of force coming from the action of the lateral pterygoids which tend to bend the mandible in the centre (du Brul & Sicher 1954). It also resists shearing forces.

inferior border forming a chin at the front. We should note that this phenomenon has taken place over a period of three million years.

Let us take the *Australopithecus Robustus* who lived three million years ago; this will be our experimental control model. How would he appear to us?

This animal-man is short (just under 4 feet) and stocky, has eyes peering out of deep-set orbits and a small skull. The face is prognathic, although concave in the centre, because the malar prominences extend forward on each side. But particularly, this far distant predecessor possessed a formidable masticatory apparatus which explains the massive size of the structures. The masseter muscle inserts on a large surface. The temporal muscle which has its origin, in part, along the mediosagittal crest is of an impressive thickness. It descends and slides through a large trough, the zygomatic arch being quite displaced (the circular appearance seen from above is impressive). He is, thus, a veritable crushing machine, the principal action of which is located in the molars and the premolars. In effect, these specimens had twice the occlusal surface of *Homo sapiens* (Fig. 4.3).

With these forces, associated with being a herbivore and the chewing of grains, an explanation is provided for the thickness of the mandible and its reinforcement in the central portion of the arch by two posterior tori. One also notes that the chin is behind the symphysis instead of being in front of it.

Above, the frontal prominence seems to be a response to the anterior occlusal forces and becomes even more noticeable when there is a postorbital frontal constriction, but this is in truth the skull and the brain which are relatively poorly developed. These latter two structures did not develop until later, doubling their volume by the stage of *Homo erectus* who evolved one million years ago. But we don't pause there. So as to better understand the differences in the structures associated with the mode of life and diet, we will progress to the *Neandertaliens* who lived about 40 000 years ago (between 80 000 and 35 000 years). This form of man, the *Ferrassie* man, knew how to hunt, produce and maintain fire and lived in small groups. Height was about 5 feet in females and 5½ feet in males. The skull, which by then had

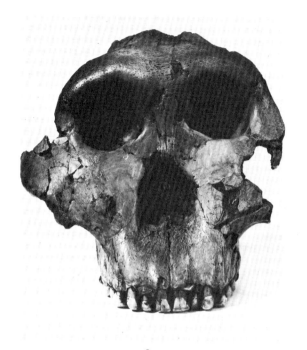

a

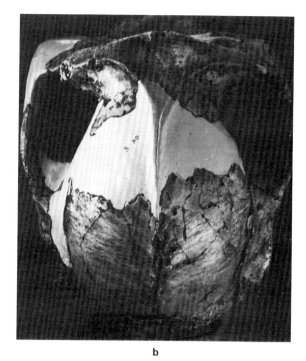

b

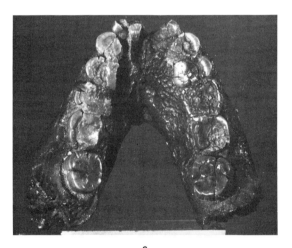

c

Fig. 4.3 **a.** Skull of Zinjanthrope (Australopithecus boisei). Olsoway 1 750 MA. Anterior view. **b.** Skull of Zinjanthrope (Australopithecus boisei). Olsoway 1 750 MA. Superior view. **c.** Mandible of *Australopithecus robustus*. Swartkrans, South Africa. 1,5 MA. (Heim 1974).

obtained a volume which would not undergo enlargement until present man, is three times larger than that of *Australopithecus*. Its surface area is large, it is broad and, a point which interests us particularly, one notes on the jaws a reduction of volume and of the large size of the molar and premolar teeth as well as a relative predominance of the incisive–canine group. This confirms the omnivorous diet with a very definite meat component. The facial architectonics translate this

change in mandibular movement: the masticatory pressures are far from being what they were in *Australopithecus*, particularly in the lower premolar group, and the supporting structures are not so thick, although in the upper jaw there are still two strong canine pillars and marked suborbital projections (Fig. 4.4).

In the centre, the nasal opening of the Neanderthal is striking by its large size.

The lower jaw is clearly less strong than it was

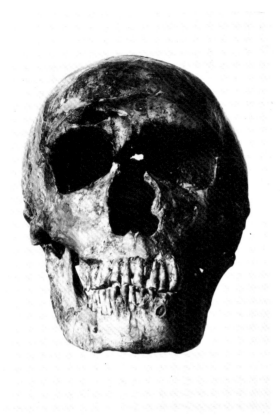

Fig. 4.4   Neanderthal: the *Ferrassie* skull. The chin is now evident in clear relief.

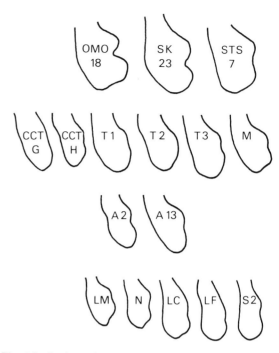

Fig. 4.5   Sections of the mental symphysis showing evolution.
First row: *Australopithecus*. Note the considerable thickness of the symphysis and the double row of tori of the mandible in each specimen.
Second row: *Homo erectus*. Attenuation of the mandibular torus (Mauer).
Third row: *Arago*. Male and female specimens, midpleistocene-Ris (Europe).
Fourth row: European Neanderthal LC = La Chapelle-aux-Saints; LF = La Ferrassie; S2 = Spy, the structure is advanced and the chin becomes outlined. It becomes even more characteristic in *Homo-sapiens* as we know him.
   Taken from Wolpoff M H 1975 Some aspects of human mandibular evolution. Determinants of mandibular form and growth. (McNamara publications, Michigan, USA).

in the *Australopithecus* and there now appears an early outline of a chin. The reduction in the dental arch also contributes to a release of the basilar border and the appearance of an outline of a chin.

In *Homo sapiens*, bringing us up to date, this latter phenomenon is encountered; the chin forms a definite projection and the face takes the shape that we have come to know since there has also been a disappearance of the suborbital fullness (Fig. 4.5).

CERVICOCEPHALIC EQUILIBRIUM

The head, atop the vertebral axis, is normally forward by its own weight, as already noted, but here there will be some necessary repetition.

Let us now examine how equilibrium is established and by which muscular action.

In profile, one can note that the head is held posteriorly by the powerful group of muscles of the nuccal region (which in the course of animal evolution with the adoption of the upright stance has become progressively disinserted, moving back down towards the base of the occiput).

In opposition to this group, we find in front a group of prevertebral muscles which act with a short lever arm. Further in front and acting with a greater lever arm, we find the chain of supra- and infrahyoid muscles in continuity with the mandible above enveloped in its pterygomasseteric sling. One must note here that the mandible is an element in this muscular chain just as the hyoid bone is below (Fig. 4.6).

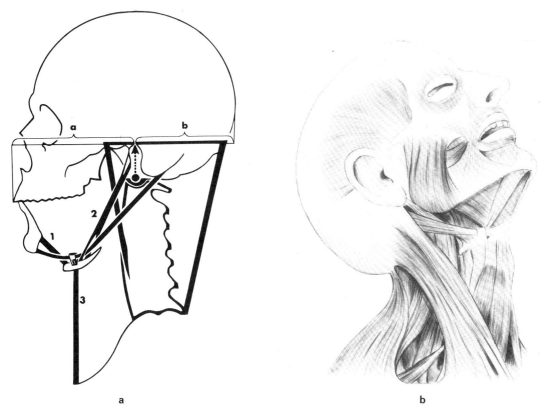

Fig. 4.6 Cervicocephalic equilibrium and stabilization of the hyoid bone. **a.** The effect of the levers and muscles. Besides the equilibrium which is established between the posterior group of muscles of the neck acting on a long lever and the prevertebral group acting on a short lever, one can see that the mandible suspends the hyoid bone and that this is held in place by the action in equilibrium of the geniohyoid, suprahyoid and stylohyoid muscles (1, 2 and 3). The digastric muscle slides freely. The anterior muscular chain parts here from the mandible, but it must be emphasized that the mandible is itself held and suspended by the pterygomasseteric sling which is a part of the anterior muscular chain. The muscular chain extends from the base of the skull to the clavicle and participates in the postural mechanism (Taken from du Brul L 1980 Oral Anatomy. C V Mosby, New York, p 201). **b.** The mandible is part of the anterior muscular chain. The postural mechanism determines its resting position, the projection of the chin which points either forward or downward and the position of the head which also depends on the 'rest position' of the mandible.

The mandible thus appears to be suspended from the skull and in turn supports the hyoid bone, and its resting position is established in response to a postural mechanism in which this entire muscular chain constantly participates.

It is easy to observe that the mandible falls naturally if there is an extension of the head posteriorly and elevates again with cervical flexion.

Seen from the front, equilibrium is assured by a system of 'stays' (nautical term) involving the scalenes and the sterno-cleido-mastoid muscles, these being the most directional (the sterno-cleido-mastoid muscles in man are relatively strong muscles and are constantly in use). From their in-

sertions on the two mastoids which are symmetrical in respect of the vertebral axis, their equilibrating action is determined by the form of the base of the skull. In other words, every type of disequilibrium will have a definite effect on the mastoid, the petrous bones and the glenoid fossae.

To return to the cervical vertebral column itself, one can see right away that it cannot be regarded as an isolated structure. It is situated at the top of a chain of curves and its degree of flexion, curvature, and posterior concavity, fixing the head according to the visual axis, obviously depends on the dorsal curve below which shapes it in response to the overall static state.

Seen in the light of animal evolution the dorsal curve is 'primitive', and in the straightening of the body, the cervical curve is a secondary, adaptive curve, placing the head in the best equilibrium, completely at the top of the spine.

In considering the relationships between the vertebrae in the area of the head, one notes that the main centre of rotational movement is located on the atlas-axis, and the centre of movement of flexion and extension is located at the hinge area but also along the entire height of the cervical column by the agency of the intervertebral discs.

THE APPEARANCE OF ANOMALIES IN RELATION TO CERTAIN POSTURAL DISEQUILIBRIA

Case 1: Congenital torticolis (Fig. 4.7)

This cervicocephalic malformation has a characteristic form. At its origin, there is sclerosis of one of the heads of the sterno-cleido-mastoid muscle which limits the rotational movements of the head. This acts on the base of the skull throughout growth and it becomes deformed. For us it is the fact that the glenoid fossae and the temporomandibular joint (which are no longer at the same level) act directly in the deviation of the chin and the mandible anomaly. The deviation is, in fact, seen as a form of craniofacial asymmetry in evolution. Delaire (1978) and his coworkers have carefully analyzed the anomaly (Fig. 4.7).

The essential postural defect has repercussions further down in a high compensatory vertebral scoliosis. In such a case, one should try to obtain freedom of movement as early as possible with removal of the fibrous tissue and the use of a plastic lengthening procedure if necessary, which was done in this case.

Case 2: Unoperated

Nose and maxillo-alveolar protrusion was associated with a retrognathia of the lower jaw, and the chin appeared to be 'swallowed' (Fig. 4.8).

The phenomenon can only be understood as a manifestation of a 'generalized' hypotonia which

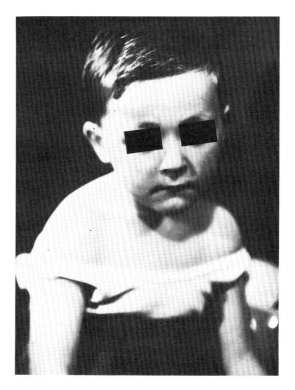

Fig. 4.7 Congenital torticollis. The cervicocephalic malformation is evident. The deviation of the chin can be seen and is part of an evolving asymmetrical craniofacial picture. The treatment is by section of the involved head of the sternocleidomastoid muscle with lengthening and physical therapy.

presented itself in infancy and led to a dorsal kyphosis and a compensatory cervical lordosis. The head tilts backwards. The chin is pulled by the superhyoid muscles (digastric and mylohyoid). The chin is 'swallowed' with posterior rotation of the mandible.

The treatment plan included:

1. orthodontic treatment with maxillary expansion;
2. an orthopaedic surgical procedure (Wassmund);
3. respiratory and other corrective exercises.

This could not be carried out. The case nevertheless remains an excellent example of this type of deformity. The diagnosis of the mandibular retrognathia in particular is meaningless if it is not placed in the entire context.

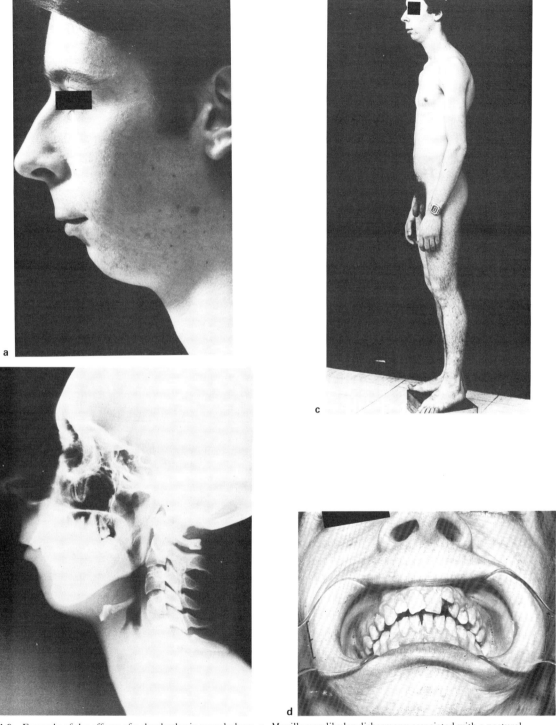

Fig. 4.8 Example of the effects of a dysrhythmic morphology. **a.** Maxillomandibular disharmony associated with a postural anomaly and hypotonia; **b.** lower retrognathia. The chin is 'swallowed'; **c.** the morphological dysrhythmia is part of the entire profile of the body.

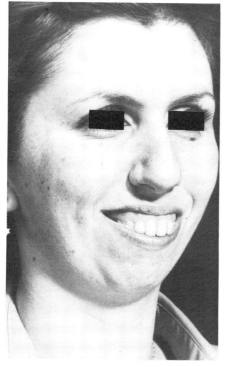

a

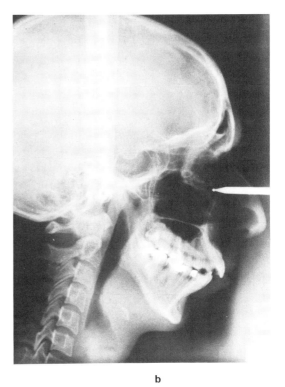

b

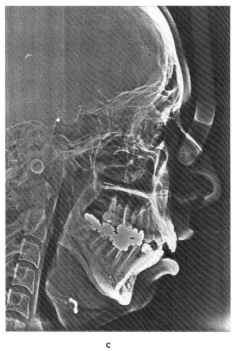

c

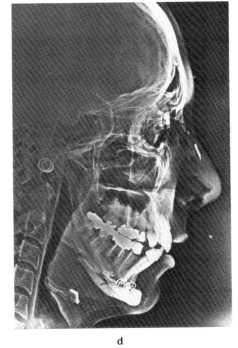

d

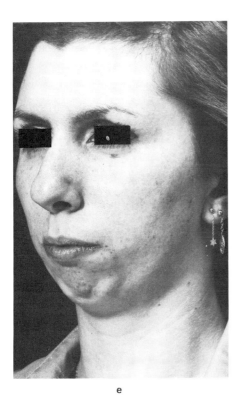
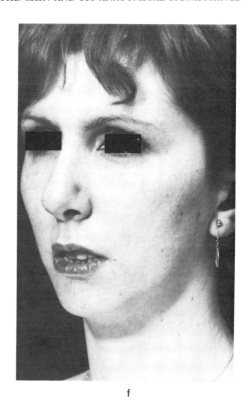

e   f

Fig. 4.9 Lower retrognathia related to a postural anomaly. Excessive protrusion of the superior alveolar segment, and nasal deformity. Correction by: a Wassmund procedure of the maxilla; advancement of the chin by basilar osteotomy; rhinoplasty.

Note that after surgery the ugly appearance caused by hyperactivity of the muscles of the crest of the chin has disappeared. Note also the change in the relationship between the hyoid bone and the inferior border of the bony chin. There has been a functional re-equilibration of the entire area (J. Lévignac & P. Doré).

Case 3

Nasal protrusion and maxillo-alveolar protrusion; characteristic retrognathia of the lower jaw; a postural defect with flat feet and accentuated vertebral curves (Fig. 4.9).

Treatment included:

1. orthodontic preparation over a period of 6 months (M. Bergé);
2. an operation of the Wassmund type with a retrusion of the incisivo-canine segment of the upper jaw and immobilization of the fragments with the use of a prefabricated splint. At the same operation a chin advancement was performed by an osteotomy and osteosynthesis;
3. rhinoplasty a year later;
4. final dental completion (C. Rousseau).

Case 4 (Fig. 4.10)

Retrognathia characterized by the complete disappearance of the chin; a postural defect with an accentuated cervical lordosis.

Treatment involved:

1. orthodontic preparation over 6 months (M. Bergé);
2. lengthening of the mandible by a procedure of the Obwegeser-Dalpont type with a 1 cm vertical lengthening;
3. intermaxillary fixation for 2 months;
4. advancement of the chin by osteotomy with sliding advancement;
5. liposuction of the neck.

The chin is freed and becomes defined in the overall cervicocephalic profile. The face is lengthened and becomes more oval.

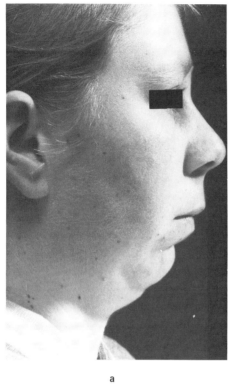

a

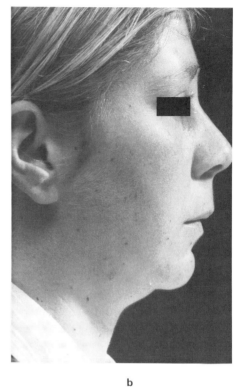

b

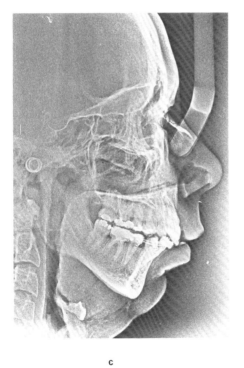

c

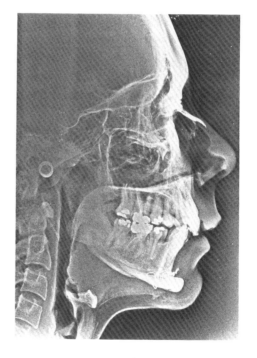

d

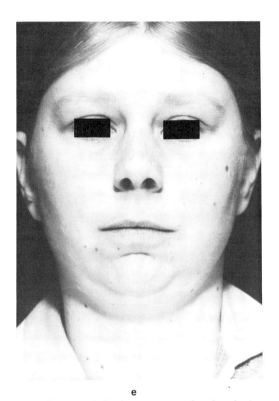
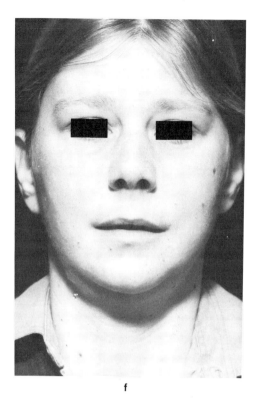

e   f

Fig. 4.10   Case: correlation between upper dorsal cyphosis, cervical compensatory lordosis, and marked mandibular retrognathia. Correction in 2 stages (J. Lévignac and P. Doré): (i) lengthening of the mandible by the Obwegeser-Dalpont technique with 2 months of intermaxillary fixation; (ii) advancement osteotomy of the chin and lipo-aspiration of the submental fat.
In the final views, note the definition and profile of the chin and the ovalization of the face in the frontal view.

Case 5: 'Long face' (Fig. 4.11)

The increase in height has an effect on the maxilla and on the mandible. The cervical lordosis compensates for a dorsal kyphosis with a tilting of the head backwards, and appears to be responsible for the development of the hyperdivergence of the face, the chin being pulled inferiorly by the muscles of the anterior chain of the neck and possibly also by the entire envelope of the platysma muscles.

The alveolar processes have thus evolved in a large open space between the two osseous bases and this is carried all the way back to the eruption and then contact of the first permanent molars, both upper and lower, which explains the excessive height according to Proffit (1978).

Treatment included:

1. primary orthodontic treatment for 6 months;
2. the Bell procedure to diminish the height of the upper jaw with intermaxillary fixation for 45 days;
3. one year later, correction of the hyposogenia (retrogenia associated with a long chin) by removal of bone, followed by chin advancement with osteosynthesis.

Case 6: Problem of ageing (Fig. 4.12)

The compression of the vertebral column with a high dorsal kyphosis calls for a compensatory tilting of the head backwards. There is a relaxation of the platysma muscles of the neck. The definition of the chin disappears in profile.

Correction is achieved by extended facelift from one side to the other and placement of a chin implant by an intra-oral approach.

It is interesting to superimpose this case on the two preceding ones. One can see that the same vertebral phenomenon, but of a different origin, can occur in infancy with a growing mandible.

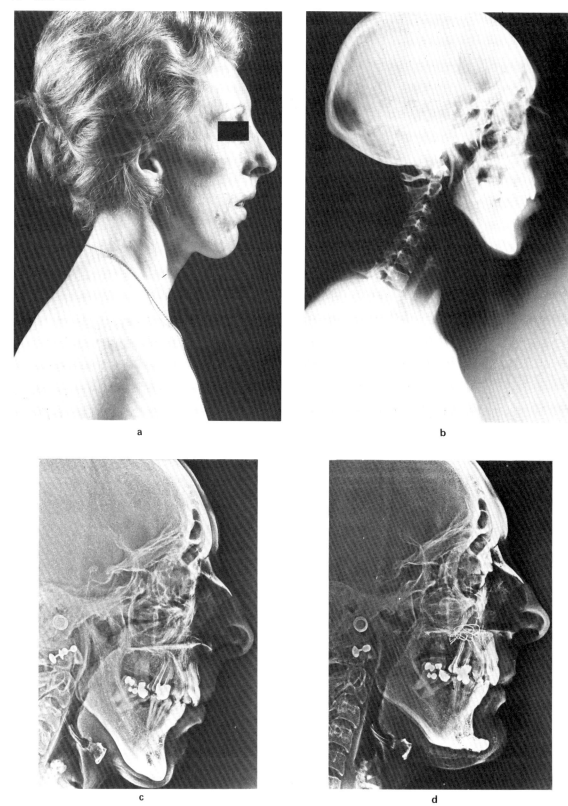

a

b

c

d

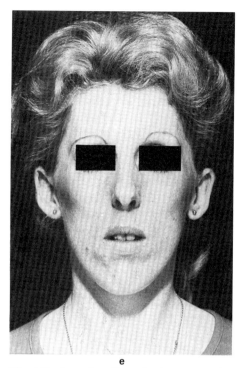
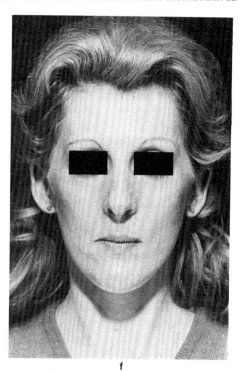

e   f

Fig. 4.11   Long face. Correction by: preoperative orthodontic treatment with maxillary expansion; calculated reduction of the height of the maxilla (Bell procedure); height reduction and advancement of the chin by osteotomy, subtraction and sliding.

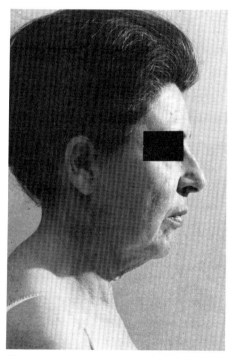
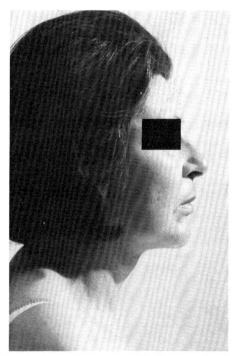

Fig. 4.12   Case of ageing. Vertebral 'settling down'. The head tilts backwards to compensate for the upper dorsal kyphosis. Relaxation of the platysma muscles of the neck accentuates the effacement of the cervicomandibular angle. Correction by facelift extended from one side to the other. Placement of a chin prosthesis by an intra-oral route.

CONCLUSIONS

The chin is an important reference point in the cervicocephalic profile. To position it properly, one must take note of the following:

1. the line of the frontal region (Muzj);
2. the projection of the nose;
3. the position of the hyoid bone;
4. the vertebral curves and cervicocephalic equilibrium;
5. the angulation at the base of the skull;
6. the degree of divergence in the development of the face, orienting the chin either anteriorly or posteriorly.

We have tried to point out the importance of postural phenomena. There are, in fact, correspondences between the planter arch, the position of the pelvis, the chain of vertebral curves, and the cervicocephalic posture. The last influences the position of the mandible and in the final analysis over a number of years defines the position of the chin in the overall cervicocephalic profile, and by the same token, the height of the face.

In the clinical cases presented, the dysmorphias are each shown to be a result of a given initial anomaly of posture and movement. Thus, if the proposed treatment takes into account the immediate problem and the dysmorphic situation, it should also look to prevention which attacks the causes from infancy. An important role should also be given to respiratory and corrective exercises.

Human equilibrium is in movement and dynamic, not static. During motion, the cervical spine supporting the head is the site of formidable postural readjustments with all of the different consequences (Desmont 1987).

This has brought us to look, for example, from the feet upwards to understand what is taking place at the head, in equilibrium on the vertebral axis. One can make better decisions with the benefit of complete understanding.

We do not want to finish this work without thanking in particular Lloyd du Brul, a wise anatomist and a friend.

BIBLIOGRAPHY

Bell W H, Creekmore T D, Alexander R G 1977 Surgical correction of the long face syndrome. American Journal of Orthodontics 71: 40–67
Björk A 1955 Cranial base development. American Journal of Orthopedics 41: 198–225
Brace L L 1977 Occlusion to the anthropological eye. The biology of occlusal development. McNamara, Michigan, pp 179–211
Cauphepe J 1955 Les variations de la tête. Orthodontie Francaise 26: 391–397
Cauhepe J 1956 Les causes de la morphogenèse. Bases de l'orthodontie. Actualites Odonto-Stomatologiques 34: 219–220
Couly G 1950 La dynamique de la croissance céphalique. Le principe de conformation organo-fonctionnelle. Actualites Odonto-Stomatologiques 117: 63–96
Delattre A, Fenart R 1960 L'hominisation du crâne. Editions du CNRS, Paris
Deffez J P 1987 Communication personnelle
Delaire J 1978 The Potential role of facial muscles in monitoring maxillary growth and morphogenesis. Muscle adaptation in the craniofacial region. McNamara, Michigan pp 157–180
Desmont G 1987 Communication personnelle (Med. Physique. Hôpital International, Université de Paris)
du Brul L E, Sicher H 1954 The adaptative chin. Charles C, Thomas, Springfield
du Brul L E 1977 Early hominid feeding mechanisms.

American Journal of Physical Anthropology 47: 305–320
Campbell T D 1925 The dentition and palate of the Australian aboriginal. Hassel, Adelaide
Fieux J 1956 Etude du comportement musculaire orofacial particulièrement pendant la déglutition, la mastication, la phonation dans ses rapports avec l'orthopédie dento-faciale. 27: 483
Gudin R G 1968 La bascule mandibulaire. Ses incidences sur le profil facial et pharyngé. S.F.O.D.F., 39: 433–449
Heim J L 1974 Les Hommes fossiles de la Ferrassie et le problème de la définition des Néandertaliens classiques. L'Anthropologie, 78(1): 81–112 and 78(2): 321–378
Kakandji I A 1982 Physiologie articulaire. Tome 3. Tronc et rachis, Maloine Editeur, Paris
Keith A 1923 Man's posture British Medical Journal Hunterian lectures 451–672
Lathan R A 1976 An appraisal of the early maxillary growth mechanism. Factors affecting the growth of the midface. McNamara, Michigan, pp 43–59
Lévignac J 1960 Le moment, les moyens, les possibilités de la chirurgie dans le traitement des difformités mandibulaires dues à un trouble de développement. Revue Francaise d'Odonto-Stomatologie 4: 491–504
Macary A F 1964 Dysmorphoses et corrélations respiratoires. Orthodontie Francaise 35: 149

Moss L M 1968 The primary of functional matrices in oro-facial growth dent. Pract. Dent. Rec., 19: 65

Mugnier A et al 1975 Génétique en stomatologie infantile. Encyclopedie Medico-Chirurgicale 220004 A10–3

Petrovic A 1975 L'ajustement occlusal: son rôle dans les processus physiologiques de contrôle de la croissance du cartilage condylien. Orthodontie Francaise 270–280

Petrovic A et al 1979 La taille définitive de la mandibule est-elle comme telle prédéterminée génétiquement? Orthodontie Francaise 50: 751–761

Proffit W R 1978 The facial musculature in its relation to the dental occlusion. Muscle adaptation in the craniofacial region. McNamara, Michigan pp 73–89

Rouvière H 1943 Anatomie humaine. Tome I. Masson, Paris, p 555, fig. 346, p 601, fig. 376.

Ricketts R M 1975 Mechanism of mandibular growth: a series of inquiries of the growth of the mandible. Determinants of mandibular form and growth. McNamara, Michigan

Sassouni V 1962 The face in five dimensions. West Virginia University Publishers, Morgan Town

Schwarz A 1926 Kopfhaltung und Kiefer. Zeitschrift für Stomatologie 24: 669–744

Solow B, Tallgren A 1976 Head posture and cranio-facial morphology. American Journal of Physical Anthropology 44: 417–436

Solow B, Kreiborg S 1977 Soft tissue stretching: a possible control factor in craniofacial morphogenesis. Scandinavian Journal of Respiration 85: 505–507

Talmant J 1979 La mandibule: un élément de la structure respiratoire ou de l'action morphologique de la mécanique ventilatoire sur la mandibule. Orthodontie Francaise 50: 671–681

Wolpoff M H 1975 Some aspects of human mandibular evolution. Determinants of mandibular form and growth. McNamara, Michigan

5. The place of the chin in the architecture of the face

J. Delaire   J. Mercier

The place of the chin, in comparison with other elements of the skull and of the face (bone and soft tissue), does not depend upon chance. It results, in fact, from the interactions which exist (from the fetal age to adulthood) between the growth potentials of all those elements and the influences of the other muscles involved in orofacial and cervical functions.

The chin is located in a position which is at the same time quite precise and also quite variable (according to subject, their facial type, their age, their sex, their race, and their oro–facial–cervical function, etc.).

A good knowledge of the phenomena which determine the position of the chin makes it possible not only to establish whether a position is normal or not, but also to determine why this is so and when necessary to work out a plan of therapy.

REVIEW OF THE PRINCIPAL MUSCULAR SYSTEMS ACTING ON CRANIOFACIAL MORPHOGENESIS (AND THEREFORE ON THE POSITION OF THE CHIN)

Three musculo-aponeurotic systems are the principal agents responsible for this morphogenesis:

1. the posterior and lateral cervical muscles;
2. the 'deep' muscles of the face;
3. the superficial musculo-aponeurotic system (SMAS) of the face.

During phylogenesis they have had an interactive effect and have led to the cephalic development which we see in man. During ontogenesis, they again model the craniofacial skeleton which does not acquire its definitive equi-

librium until the acquisition of the erect posture of the head.

I. The posterior and lateral cervical muscles

Their mass, their strength and the extent of their insertions on the occipital expanse as well as the mastoid processes are considerable. In addition, their modelling action extends to the superficial portions of the top of the skull through the epicranial aponeurosis and the scalp, and deeply by way of the intracranial meningeal aponeurosis (the falx of the brain, the tentorium of the cerebellum) (Fig 5.l).

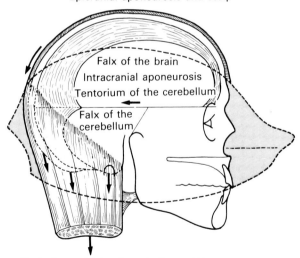

Epicranial aponeurosis and scalp

Falx of the brain
Intracranial aponeurosis
Tentorium of the cerebellum
Falx of the cerebellum

Posterior and lateral cervical muscles

Fig. 5.1 Action of the posterior and lateral cervical muscles on the occipital expanse, the petrous bones, the cranial vault (by the epicranial aponeurosis and the falx of the brain and the cerebellum) and the base of the skull (by the cerebellar tentorium). The sternomaxillary band joining the sternocleidomastoid to the angle of the mandible contributes to the morphogenesis of this bone.

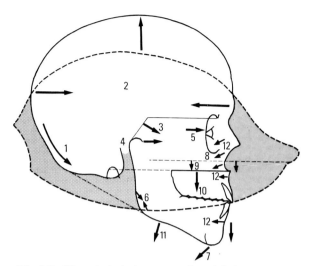

Fig. 5.2 The principal phenomena of craniofacial hominization.
1. posterior occipitotemporal rotation;
2. anteroposterior reduction of the skull, with increase in its height and length;
3. angulation of the base of the skull;
4. transverse orientation of the petrous pyramids with anterior migration of the temporomandibular joints;
5. frontalization of the orbits;
6. bending of the mandibular angle;
7. recession and lowering of the chin;
8. verticalization and lengthening of the maxilla;
9. lowering of the palatal plane;
10. lowering of the occlusal plane;
11. lowering of the basilar border of the mandible;
12. remodelling of the anterior portions of the facial skeleton with formation of paranasal hollows and the recession of the dental arches in respect to their osseous bases (with a prominence of the nose, the nasal spine and the cheekbone area).

They are thus the principal agents for the development of the human type of skull (Fig 5.2), notably:

a. the posterior occipitotemporal rotation;
b. the anteroposterior reduction of the skull with an increase in its height and width;
c. the angulation of the cranial base;
d. the transverse orientation of the petrous pyramids with the anterior migration of the temperomandibular joints;
e. the frontalization of the orbits.

To these cranial modifications should be added the transformations of the facial skeleton which also characterize the process of development of the human skull: the bending of the angle and the opening of the mandibular curve, the progression and downward displacement of the chin, the verticalization and lengthening of the maxilla and the depression of the nasal spine, the palatal plane, the occlusal plane and the basilar border of the mandible.

All of these cranial and facial phenomena are accomplished at the same time and in close correlation. One can then understand the reason for the parallelism of the anterior portion of the base of the skull and of the hard palate, for its alignment on the occipitospinal joint articulation, and for the basilar border of the mandible on the inferior protrusion of the occipital expanse.

2. The 'deep' muscles of the face

These are organized in a complex ensemble of muscles and participate in the formation and the movements of the tongue, the soft palate, the floor of the mouth and the lateral walls of the pharynx.

In animals, whose posture is horizontal, the thoracoabdominal viscera are completely suspended from the thoracic rib cage and the abdominal wall. The floor of the mouth supports only the tongue and as a consequence has very little muscle.

In man, to the contrary, a large portion of the weight of these viscera is supported by the hyoid apparatus. This results, therefore, in an important descent of this structure and of the body of the mandible (to which it is attached by the mylo- and geniohyoid muscles) (Fig. 5.3).

In accordance with this, the lingual mass tilts posteriorly, thus exerting an endobuccal aspiration which causes a recession and lowering of the dento-alveolar arches and the inferior border of the upper maxilla.

The tractions exerted by the genioglossus, geniohyoid and mylohyoid muscles on the inner surfaces of the anterior extremities of each hemimandible cause, on the other hand, a compression of the symphysis (Fig. 5.4). This occurs through the contractions of the lateral pterygoid muscles which bring together each hemimandible, more noticeable as the condyles separate further (as a consequence of the change in orientation of the petrous bones (see above)). The end result is an ossification of the symphysis cartilage and a

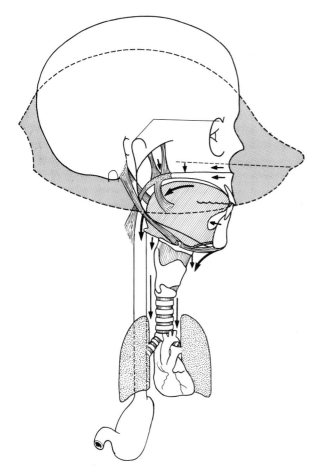

Fig. 5.3 The weight of the thoraco-abdominal viscera pulls on the hyoid complex, lowering and pulling back the body of the mandible, which tilts the tongue backwards, and also involves the regression of the alveolodental arches (on their osseous bases) and a recession and lowering of the inferior portion of the maxilla.

reinforcement of the chin occurring as the infant lifts his head and begins to walk.

Thus, the diverse actions of the deep muscles of the face contribute directly to the phenomenon of the development of the human face (see Fig. 5.2). They explain also the close correlations which exist between:

a. the orientation of the maxilla and the position of the bony chin;

b. the relative heights of the superior and inferior portions of the face and, on the other hand, of the superior portion of the face and the ascending rami of the mandible.

3. The superficial musculo-aponeurotic system of the face (SMAS)

This is largely formed by the 'platysma' or 'skin' muscles of the face (Fig. 5.5). Their morphogenetic action is exerted in particular from the inferior border of the orbits to the body of the mandible and laterally to the high points of the malar bones. From above to below, they form, schematically, three interconnecting rings.

a. The first ring (nasal) is formed by the elevator muscles (superficial and deep) of the upper lip and of the nasal alae and the transverse muscles of the nose (constrictors of the nostrils). These muscles provide both movement of the nostrils and support of the upper lip. Note that the insertions of the transverse muscles onto the incisive crest contribute to the establishment of the anterior nasal spine.

b. The second ring (labial) is formed by the superior and inferior obicularis muscles (and, to a certain extent, by the deeper muscles which, according to some authors, are simply expansions in the incisive region of the transverse muscles of the nose). They assure the movements of projection and compression of the lips (against each another and against the underlying dento-alveolar structure).

c. The third ring (mental) is formed on each side by the triangularis muscle of the lips, the quadratus muscle of the chin and the muscle of the mental crest. The triangular muscles participate in the control of the vertical and transverse position of the chin (according to certain anatomists, the triangular and oblique portion of the superior obicularis form a single muscle extending on each side from the nasal spine to the bony chin).

The quadratus muscles of the chin and of the mental crest attach the lower lip to the mental eminence and the cutaneous portion of the chin to the incisive alveolar area of the lower jaw. Normally, a relationship exists between these different elements such that, in a resting position:

(i) the lips are spontaneously in contact;
(ii) the cutaneous portion of the chin exactly overlies the bony chin (without any need for muscular activity);
(iii) the lower lip, supported (on its deep

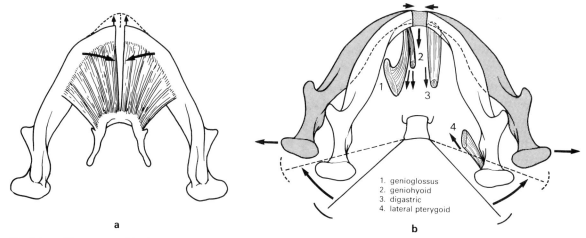

Fig. 5.4 **a.** Forces exerted by the mylohyoid muscle on the internal surface of the two hemimandibles presses the anterior extremities of these against one another. An ossification of the symphysis and an augmentation of the mental bone results.
b. Tractions exerted by the genioglossus, geniohyoid and digastric muscles on the internal slope of the anterior extremities of the two hemimandibles, accentuating the compression of the symphysis and, on the other hand, moving it backwards, contribute to the opening of the mandibular curve. The contractions of the lateral pterygoid muscles accentuate this compression which is strengthened even further as the condyles are displaced forward and downward under the influence of the modifications in the orientation of the petrous bone (homonization of the cranial base).

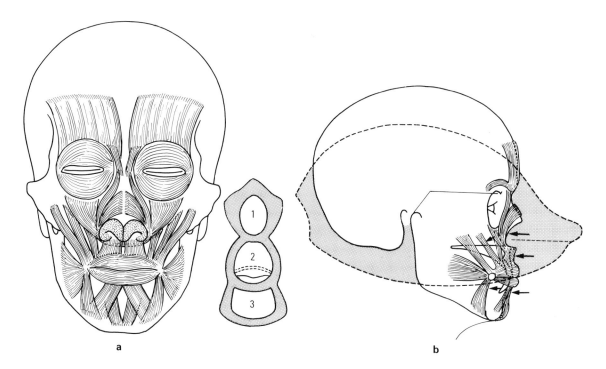

Fig. 5.5 From the infra-orbital region to the inferior border of the mandible, the muscles of the skin of the face form three rings schematically: 1. nasal — superficial elevators and deep muscles of the upper lip and of the nasal alae, and the transverse muscles of the nose; 2. buccal — obicularis muscles of the lips, fan-shaped muscles; 3. chin — triangularis muscles of the lips, quadratus muscles of the chin and muscles of the crest of the chin. Onto the external portions of the labial ring also insert the canine, zygomatic and buccinator muscles.

surface) by both the edge of the upper incisive teeth and by the mental eminence, curves harmoniously between these two supports. The result is a good concavity between the alveolar process and the bony chin (called the concavity of point B) and a good orientation of the incisors one to another and to their respective osseous bases (Fig. 5.6).

These three rings (nasal, labial and mental) and the canine, zygomatic, and buccinator muscles which insert on the outer slope of the labial ring, exert pressure on the underlying skeleton which adds to the suction created by the posterior movement of the tongue (see above). This pressure actively contributes to modelling resorption in the paranasal maxillary regions and the mandibular body, and inversely contributes to the projection of the nasal tip, the nasal spine and the bony chin.

To recap: the muscles of the neck, the velolingual complex, the buccal floor and the superficial platysmal muscles participate in the displacement and modelling of elements of the craniofacial skeleton. During the process of human cephalic formation, all these muscles have a close intercor-

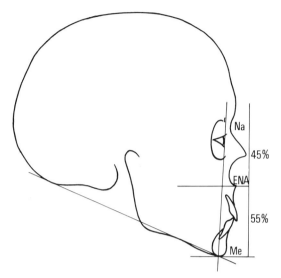

Fig. 5.7 The bony chin point (Me) is normally situated at the convergence of the anterior and occipitomandibular lines of craniofacial equilibrium, at a level such that the upper and lower portions of the face (the measurements between the levels Na–ENA and ENA–Me) measure respectively 45% and 55% of the total height of the face.

relation. This is also the case with the posterior occipital-temporal rotation, the advancement of the condyles, the lowering and regression of the mandibular body and the verticalization and lengthening of the maxilla. These factors explain why there is equilibrium and proportion between all these craniofacial elements regulating their mutual positions. Thus, normally, the osseous part of the chin is situated exactly at the convergence of the anterior and occipitomandibular lines of craniofacial equilibrium at a level determined by the normal proportions of the superior and inferior portions of the face (respectively 45% and 55% of the total height) (Fig. 5.7).

OTHER FACTORS REGULATING CRANIOFACIAL MORPHOGENESIS. THEIR INFLUENCE ON THE POSITION OF THE CHIN

Craniofacial morphogenesis depends not only on functional influences, but also on the *growth potential* of all the tissues and organs of the head: the brain, the eyes, the volume of muscles, the salivary glands, the fatty tissue, bone, cartilage and even the teeth.

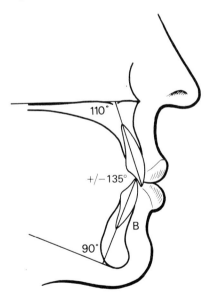

Fig. 5.6 The lower lip is normally supported along its posterior surface both by the edge of the upper incisors and the mental eminence, curving harmoniously between these two supports. Thus, a pleasant concavity at point B and a good orientation of the incisors (between themselves and in relation to their osseous bases) results.

All the soft tissues and certain skeletal elements have, in effect, a volume and essential primary growth capacity capable of influencing the displacement, the orientation and the form of the osseous craniofacial structures.

The mandible depends particularly on the volume growth potential of the tongue, the labiomental muscles, the teeth and the alveolar processes. The growth 'potential' of the condylar cartilages also plays an important role in the normal variations of mandibular growth and, therefore, in the position of the chin. In effect, in spite of the 'secondary' nature of these cartilages (demonstrated notably by Petrovic) their growth does not depend only on the functional influences exerted on them but also:

1. on their original dimensions (doubtless linked to the cells of the neural crest from which they came);

2. on general influences, notably hormonal. Thus a large condyle, in which there is a significant amount of cartilage, grows more than a small condyle under the influence of the same functional solicitations. At the time of puberty, its growth spurt is also clearly more substantial than that of a smaller condyle.

The quantity of primary growth of the chondrocranium, notably the nasal septum, varies also according to its dimensions (volume and proportions). The superior base of maxillary implantation (the frontomaxillary articulation) can also be situated in a varying forward position, i.e. at a varying distance from the temporomandibular joint (TMJ) (upon which depends the position of the mandible, and therefore the chin, in relation to the maxilla).

Taking account of the great individual variations in the growth potential of these various organs and tissues one can understand the importance of necessary adaptations in order for the facial skeleton to properly equilibrate itself and, in particular, for the chin to be located in the best functional position.

THE 'NORMAL' POSITION OF THE CHIN

We have observed above that the osseous chin has a position dictated by skeletal equilibria (anterior facial and occipitomandibular) and that the soft tissues situated in front of it are normally supple, relaxed and have good bilabial contact.

This is well objectified by clinical examination and by supplementary analysis.

I. Clinical examination

a. Frontal view. The chin is normally median, symmetrical and situated at a distance from the subnasal point equal to that which separates it from the nasal frontal line.

b. The profile. One can note the equality of the superior and inferior portions of the face. On the other hand, the normal labiomental profile is harmonious (as objectified by the 'aesthetic' lines of Steinert and Ricketts). Depending on the *facial type* the chin can, nevertheless, have a varying position relative to the frontal region. There are, in effect, 'normal' orthofrontal, transfrontal and cis-frontal types with, in each of these cases, a good labial contact, and without any excessive labiomental muscular activity.

2. Supplementary examination

a. The profile. Craniofacial architectural analysis of a subject of the white race (Fig. 5.8) shows that normally the point of the bony chin (Me) (union of the posterior curve of the symphysis and its basilar border) is situated exactly at the junction of CFl and CF6, at the lower extreme of CF5. This very precise localization of the normal chin point does not, however, exclude considerable individual variation.

In effect:

(i) *In a vertical direction.* In the infant, the inferior portion of the face is often a bit larger (about 57% of CF5). This variation (of 2%) can also be observed in white adults. In blacks, the lower portion of the face is normally much greater and can attain or even exceed 60% of CF5.

In all cases, the observation of good bilabial contact without any muscular effort makes it possible to establish a good vertical position of the chin, *provided that there is no anteroposterior anomaly*. In this situation, the placement of the bony chin in a better, more anterior position can modify its vertical position as well and therefore show up

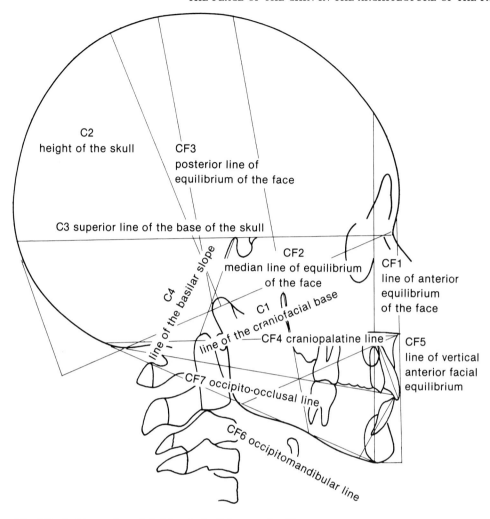

Fig. 5.8   Architectural craniofacial analysis of the normal subject.

a vertical anomaly. We will see later how one can objectify these vertical anomalies camouflaged by a recessive chin.

(ii) In the anteroposterior direction. The position of the chin varies considerably according to subject, age, facial type, sex, etc. which can all be completely normal given the condition that points FM, NP and Me are well-aligned on CF1. The angle C3-CF1 (angulation of the anterior line of equilibrium of the face in relation to the superior surface of the base of the skull) varies depending on a number of factors. Depending on age, it can increase by 5 degrees and even more between infancy and adulthood; depending on sex, it is habitually more pronounced in the male than in the female; and variation occurs according to the dimensions (length) of the base of the skull and the value of its anterior and posterior angles. In subjects with dentofacial dysmorphoses, these variations in the dimension and angulation of the base of the skull and the orientation of CF1 determine where the 'ideal' anteroposterior position of the bony chin lies.

b. Frontal view. On the frontal cephalometric film, the bony chin is normally in the midline and symmetrical in relation to the midline of the head. When this midline is difficult to establish (facial scoliosis), a vertical cephalometric technique is used to distinguish anomalies of the mandible from those of the base of the skull.

ABNORMAL POSITIONS OF THE CHIN.
METHODS OF OBJECTIVE DESCRIPTION

The position of the chin can be abnormal in all directions (vertically, anteroposteriorly and transversely), and be associated or not associated with occlusal abnormalities and with more distant skeletal anomalies (which can have a definite effect on the anomaly of the chin).

Anomalies in the position of the chin should, therefore, be carefully analyzed in order to determine exactly which skeletal anomalies are responsible.

I. In the more simple cases a bad chin position results from an anomaly of the bony chin itself. Other skeletal abnormalities are not present. Surgical correction is, therefore, performed on the chin itself (genioplasty).

Example No. 1: A patient, 16 years of age, came to see us on 27 May 1980 for an excessive inferior height of the face, a habitual open lip position and a persistent lack of incisive contact in spite of orthodontic treatment (Fig. 5.9).

Cephalometric analysis of the craniofacial architecture (Fig. 5.10) showed an excessive height of the bony chin (of approximately 7 mm) and a retroposition (of approximately 1 cm). However, the occlusal plane was well-positioned. There was also a slight palatal version of the upper incisors and canines (due to orthodontic treatment) and early alveolar resorption around the lower incisors.

A structural study (examination of a cephalometric film taken with the teeth held together but the lips in a position of repose) found the upper incisive borders were at a good level in relation to the lower lip although there was an absence of contact between the lips. A flattening of the anterior surface of the bony chin could also be observed.

To sum up, the anomaly of the position of the bony chin was both vertical and posterior, associated with a bilabial opening, but without a vertical and anteroposterior maxillary anomaly. A vertical geniectomy of 7 mm with an advancement of the chin by 1 cm was indicated. The technique used was that of Michelet, an advantage of which is the elevation of the bony chin and the muscles which insert on its anterior surface.

The postoperative results conformed to the preoperative plan (Figs. 5.11 and 5.12), and the

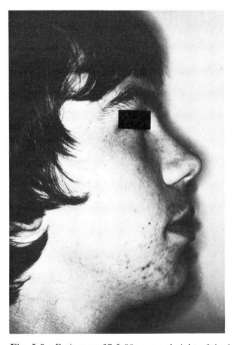
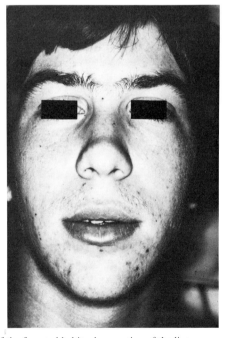

Fig. 5.9   Patient on 27.5.80: excess height of the lower third of the face and habitual separation of the lips.

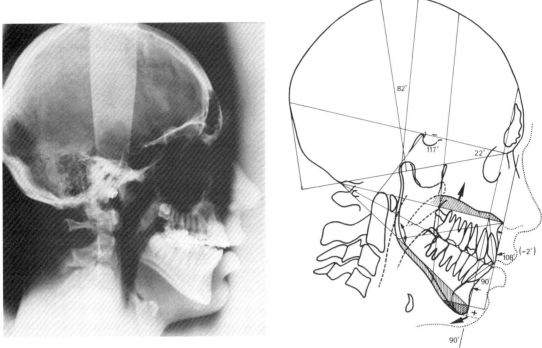

Fig. 5.10 A cephalometric film taken of the patient on 27.5.80 and its architectural analysis showing the excess height of the bony chin (approximately 7 mm) and its recession (about 1 cm). On the other hand the occlusal plane is well-situated. Note also the good level of the border of the upper incisors with respect to the upper lip, the absence of bilabial contact and the flattening of the anterior surface of the bony chin.

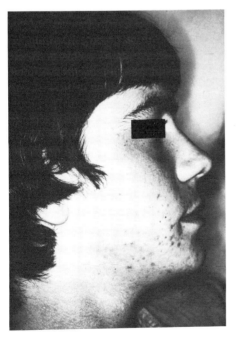
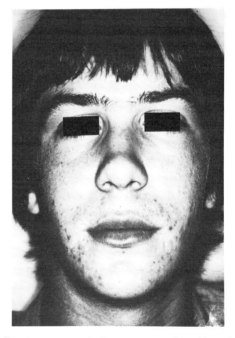

Fig. 5.11 Patient on 8.9.80, 3 months after genioplasty. Note the improvement in the appearance of the chin and the good bilabial contact.

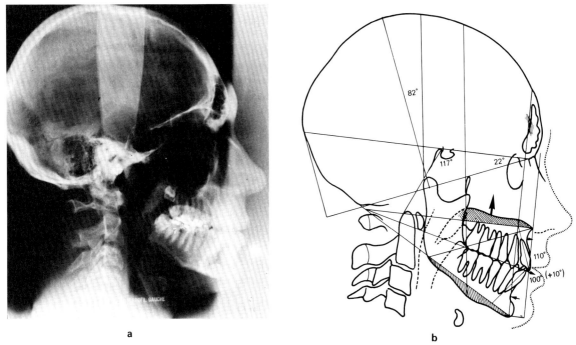

Fig. 5.12 Cephalometric film of patient on 4.3.81 (ten months after the operation) and its architectural analysis. Note that the bony chin is now in a good position in both the anteroposterior and vertical planes.

placement in a good anteroposterior and vertical position of the bony chin brought with it good bilabial contact in the rest position. (Note that the change in orientation of the lower incisors in relation to the basiliar border depended only on operative alterations of this lower border; the improvement in orientation of the upper incisors and the disappearance of the small lack of incisive contact occurred spontaneously.)

2. In other cases, some of the abnormal positions of the chin do not result from an anomaly of the chin itself but from another mandibular anomaly. This can consist, for example, of an excessive mandibular opening with an advancement and lowering of the bony chin.

In the analysis of the craniofacial architecture it is thus necessary to add certain measures which indicate whether or not, in addition to the anomalies of the body of the mandible, there are abnormalities of the bony chin. To do this one measures the distance from the lower incisive border (i) to the point of the bony chin (Me) and one compares this to half of the segment ENA to the level of Met(x), to which one adds 4 to 5 mm (in

order to take account of the root of the incisor and the fact that the segment i to Me is oblique (Fig.5.13)). In practice, to know if the point of the bony chin is at a good distance from the lower incisive border one takes half of the distance between ENA and level Met, adds 4 to 5 mm

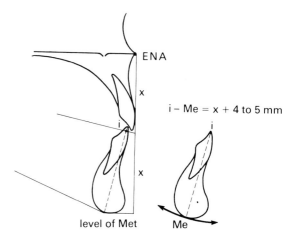

Fig. 5.13 i to Me is normally equal to x (half of ENA to the middle of Met) + 4 to 5 mm.

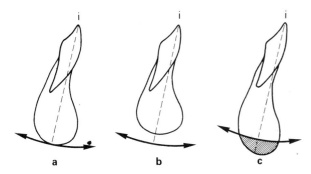

Fig. 5.14 The Arc of the circle traced from i, with a radius equal to x + 4 to 5 mm, is normally tangential to the inferior border of the bony chin (**a**). If it passes below the chin (**b**) this is too short (or there is an inferior incisive intrusion). The chin is too long if the arc of the circle cuts the chin (**c**).

(depending on the age of the subject) and using this measurements as the radius draws an arc of a circle from the point of the lower incisor (Fig. 5.14).

The arc of the circle can then be either tangental to the lower portion of the symphysis of the chin

(height of the chin is normal), pass below this (height of the chin is insufficient) or pass above its inferior border (chin is too long). One must always take account of the curve of Spee in this procedure in order to correct those situations where the lower incisive borders are too high or too low, associated with an extrusion or an intrusion of the teeth.

Example No 2: A patient, 18 years of age, consulted us on 13 October, 1980 concerning the correction of a mandibular prognathism with a significant anterior open bite (Fig. 5.15). In rest position, there was no contact of the lips and the chin was located too low and too anteriorly.

Analysis of the craniofacial architecture (Fig. 5.16) confirmed the abnormal position of the chin located both too anteriorly (approximately 8 mm) and too low (approximately 10 mm). The height of the chin (lower incisive border to the point of the bony chin) was in the normal range (38 mm, or only +2 mm greater than the ideal measurement of 36 mm). The inferior premolar and molar occlusal plane was clearly oblique in an inferior direction while the occlusal plane of the superior molars and premolars was at a good level. The

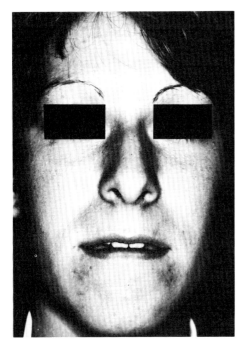

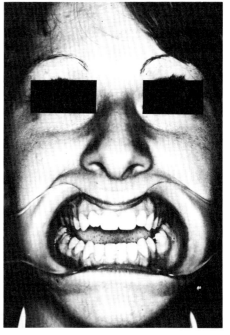

Fig. 5.15 A patient, 18 years of age, on 13.10.80. At rest there exists a bilabial separation. Note the macrogenia and the malocclusion.

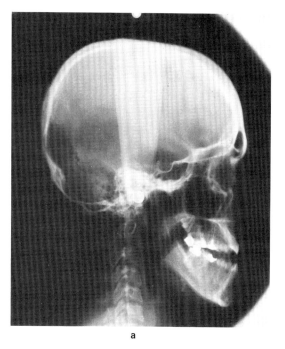

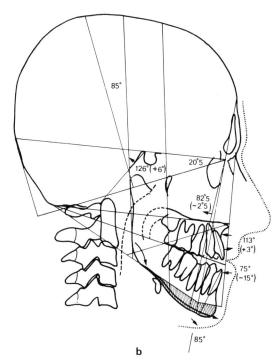

Fig. 5.16 The cephalometric film of the patient on 13.10.80 and its architectural analysis show that the anomaly in the position of the chin results essentially from the opening of the mandibular angle; on the other hand, the length of the segment of the inferior incisive point of the bony chin is subnormal. The excessive opening of the posterior angle of the cranial base favours a cis-frontal type. In spite of the slight retromaxillar anomaly and the superior retro-alveolar anomaly, a sagittal osteotomy of the ascending rami was decided upon.

border of the upper incisors was 9 mm from the cingulum of the upper incisors. There was, in addition, a clear opening of the mandibular angle.

Moreover, the maxilla was slightly tilted posteriorly, but the posterior angle at the base of the skull was clearly greater than normal (+6 degrees) which predisposes to the cis-frontal type.

To sum up, the dentofacial dysmorphosis was essentially due to the excess opening of the mandibular angle.

Study of the dental models and the surgical simulation on the cephalometric film showed again that the two dental arcades were concordant after closure of the angle. The bony chin came into a good position.

A sagittal osteotomy of the ascending rami of the mandible (Dalpont/Epker technique) was chosen as a consequence which gave the anticipated good results.

Eight months after the operation (Fig. 5.17) the labiomental relationships were satisfactory and the occlusion remained stable. Architectural cranio-

facial analysis showed a good anteroposterior and vertical position of the chin (Fig. 5.18).

3. In other cases where there is an anomaly of maxillary height, the anomaly of the chin can then be due uniquely to the maxillary abnormality or to a combination of abnormalities of the maxilla and the mandible. Very complex cases are seen which necessitate a very precise diagnosis of the location and extent of the various anomalies. To do this one combines architectural craniofacial analysis with the measurement of the ideal height of the osseous chin (see example No. 2) and other procedures which make it possible to see the effect of subsequent maxillary elevation on the position of the bony chin.

The extent of elevation of the maxilla is itself appreciated not only by architectural analysis, but also (in clinical practice and on the cephalometric profile film) by the measurement of separation (in a rest position of the lips) between the free border of the upper lip and the border of the upper incisors.

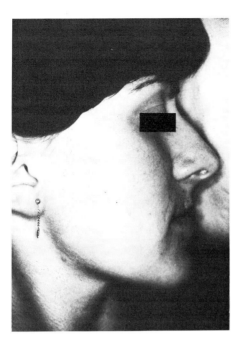
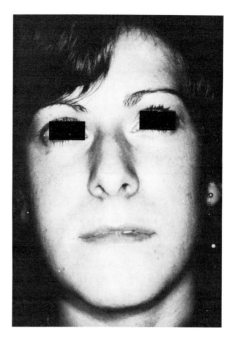

Fig. 5.17 Patient on 10.6.81, six months after the operation. The height of the inferior portion of the face is normal and there is a good bilabial contact.

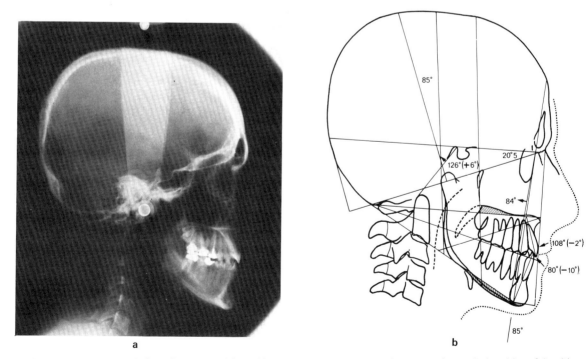

Fig. 5.18 The cephalometric film of 10.6.81 and its architectural analysis show a good postoperative vertical position of the chin.

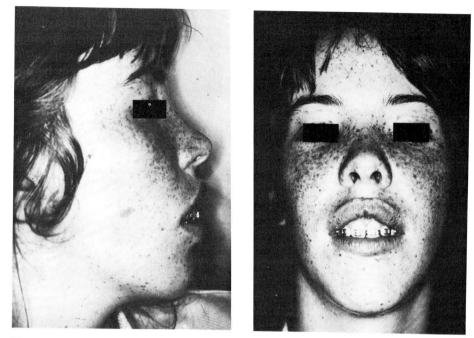

Fig. 5.19   A patient, 18½ years of age, on 1.9.80. Note the extensive lack of labial occlusion and show of the upper incisors.

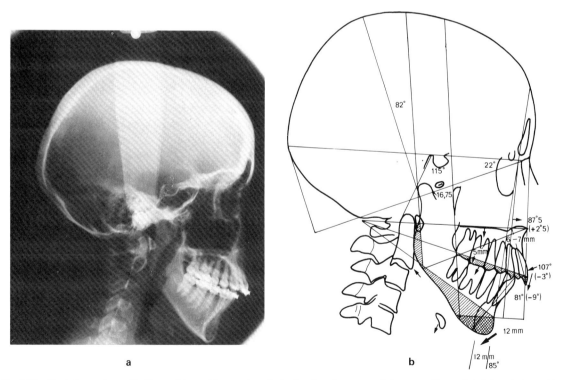

a   b

Fig. 5.20   Cephalometric profile film and its architectural analysis showing the extent of the anomalies and defining their positions on the maxilla and the mandible: a promaxillar anomaly, vertical maxillary excess, lowering of the occlusal plane, posterior tilting of the ascending rami of the mandible, lowering of the bony chin by approximately 12 mm with a retrusion of an equivalent amount.

Example No. 3: A patient, 18½ years of age, came to see us on 1 September 1980 about an excessive height of the lower portion of the face causing an unaesthetic facial appearance and a habitual lack of occlusion of the lips (Fig. 5.19). There was also a significant anterior projection of the upper incisors in relationship to the lower incisors.

Craniofacial architectural analysis (Fig. 5.20) showed the complexity of the dentofacial dysmorphosis caused by the association of a maxillary protrusion (+2° 5), an excess vertical height of the maxilla (with a lowering of the occlusal plane of the molars by approximately 5 mm and of the canines by 6 to 7 mm), a posterior movement of the ascending rami of the mandible and a downward displacement of the bony chin level with a regression of 12 mm.

Half of the ENA to Met level = 34 mm. The length of i to Me should therefore be 34 + 5 = 39 mm. It, in fact, measured 49 mm, showing an excess of 10 mm (Fig. 5.21).

The study of the effect of raising the upper molars (by 5 mm) on mandibular rotation and the position of the upper incisors of the upper lip (Fig. 5.22) showed that this elevation was not adequate to correct the alteration existing between the free border of the lip and the upper incisors. In effect, there remained a difference of 5 to 6 mm. It was, therefore, necessary to increase the molar elevation by another 3 mm (associated with a set back of the incisive–canine block to avoid excessive maxillary

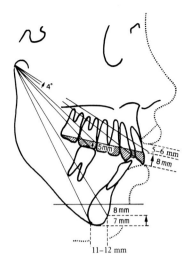

Fig. 5.22 The 'ideal' positioning of the occlusal plane (determined by architectural analysis: elevation of 5 mm at the level of the first molars) requires a mandibular forward tilting of 4 degrees. This movement involves an elevation of the edge of the upper incisors of 8 mm which is inadequate when one considers the separation between the free border of the upper lip and the incisive border (13 to 14 mm). A further elevation of the molars (by 2 to 3 mm) was, therefore, necessary.

The tilting of the mandible was 7 degrees. This involved an elevation of the chin by approximately 7 mm in advancement of 11 to 12 mm. A complementary geniectomy of approximately 8 mm was therefore adequate for the bony chin to end up at a good level.

projection). The mandibular rotation achieved came to 7 mm and elevated the bony chin by 7 mm with an advancement of about 11 to 12 mm.

The following operation was therefore decided upon: a four-segment maxillary osteotomy with an extraction of the two maxillary bicuspids and an elevation of the molars by approximately 1 cm; a reduction genioplasty, shortening the height of the bony chin by about 8 mm. The operation gave the anticipated results.

Six months after the operation, facial equilibrium was satisfactory with good bilabial contact and a good aesthetic result (Fig. 5.23). Architectural analysis (Fig. 5.24) shows that the result was obtained at the cost of a slight elevation of the occlusal plane, which was well-tolerated by the patient.

Thus, thanks to craniofacial architectural analysis and other types of tracings, it is possible to establish precisely the nature of anomalies of the position of the chin and to decide upon indicated

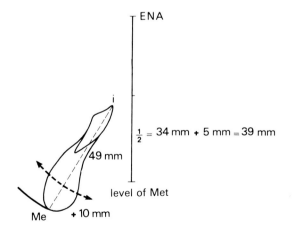

Fig. 5.21 The segment i to Me exceeds its normal length by 10 mm: ½ of ENA to level of Met + 5 mm = 39 mm, whereas i to Me = 49 mm.

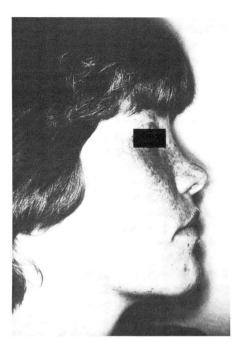
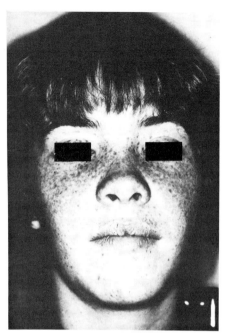

Fig. 5.23   The patient 6 months after surgical intervention.

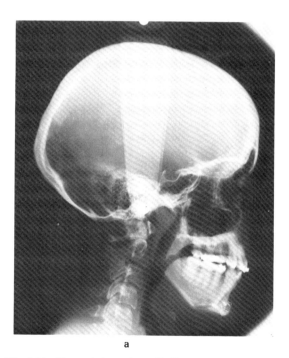

a

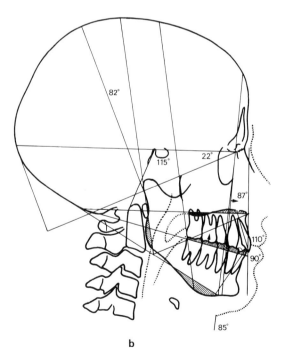

b

Fig. 5.24   The cephalometric profile film and its architectural analysis show the good anteroposterior and vertical position of the bony chin and the good relationships between the lips themselves and between these and the incisors. This result was obtained at the price of a slightly excessive elevation of the occlusal plane.

therapy (specifically surgical). In addition to the cephalametric studies, it is important to carry out a careful examination of the soft tissues of the chin since the information provided by clinical examination may alter the operative technique somewhat.

CONCLUSIONS

The place of the chin in relation to the other elements of the skull and the face depends on the interactions which exist between the growth potentials of all these elements and the influence placed on them by the muscular systems of the face and neck: the posterior and lateral occipital muscles, the deep muscles of the face and the superficial facial muscles.

The different skeletal phenomena of craniofacial 'hominization' which are the result of these influences, have close correlations which are well-objectified by craniofacial architectural analysis.

In white people, the lower portion of the chin is normally located at the point of convergence of two lines of equilibrium: the 'anterior' line of the face and the 'occipitomandibular' line, in this analysis. It is located also at a level determined by the well-defined proportions which normally exist between the upper and lower portions of the face (45% and 55% of the total facial height).

Other complementary tracings help in determining the ideal position of the bony chin in cases where there are significant occlusal abnormalities (and when maxillary surgery is necessary to modify the level of the occlusal plane).

Particular attention should also be paid to the soft tissues, and in particular, to the relationships existing between the lips, the border of the upper lip and the upper incisors, the level of the cutaneous chin and the bony chin, and to whether there are straining contractions in the mental musculature. A good position of the bony chin should be associated with good relationships, both in rest and in function, of all of these elements.

6. Morphologic anomalies of the chin (affecting the soft tissue)

P.-A. Diner   P. Oxeda   J.-M. Vaillant

The evolution of anomalies in the form of the chin should be considered in terms of real and relative morphology. Real morphology depends on a number of characteristics both at the level of the soft tissues and the bony symphyseal support. Bony support, by its position (pro- or retrogenia, laterodeviation), its height and its shape modifies the drape of the soft tissues.

The soft tissues of the chin are difficult to separate from those of the neighbouring labial or submental areas. One should examine the labiomental region (the 'lower lip–chin complex' as described by Bell) rather than the chin as an isolated element since it exists in a permanent and reciprocal relationship between the various regions and since the analysis of one region involved comparison to another region.

Finally, the chin is integrated into the overall face and it is necessary for the evaluation of the chin to be done in terms of function and in relation to the other morphological elements of the face (facial harmony and equilibrium).

FUNDAMENTAL PRINCIPLES

1. The mental skeleton (aside from anomalies of its position)

The proportions and the profile of the mental osseous protruberance are extremely variable from one subject to another, from one ethnic group to another, and in the same individual as a result of his age, e.g. the absence of a mental eminence at birth.

Certain authors have attempted to relate the form of the protruberance to the morphology of the rest of the mandible (Haskell), e.g. diminution of the protrusive character in cases of vertical development of the mandible.

In fact, there is an extreme variability in form which takes account of the reciprocal influence of the soft tissues and bony tissues during growth and the intrinsic genetic character of the mandible (cf. Goret-Nicaise, Ginisty), and the hereditary and functional duality which is constant during craniofacial growth.

Certain surface characteristics stand out as having more projection or flattening: the curve of the anterior mandible, the depression located above the mental eminence and the confluence of the mental fossae; the central symphyseal tubercle of the trigonum whose base runs between the lateral and marginal tubercle which forms a medial crest and the anterior bend of the basilar border of the symphysis.

2. Tissue layers covering the chin

These are mobile and include the periosteum, the muscular layer with the insertions of the quadratus of the chin, the triangularis of the lips, inserting on the external oblique line of the mandible on the lower lip, and the homolateral commissure, the muscle of the mental crest, inserting on the mental fossae, and inferiorly on the skin of the chin, the fatty subcutaneous tissue traversed by muscular fibres going to the particularly thick skin of the chin, which occasionally has small cushions of fat.

The thickness of the soft tissue of the chin is on average 12 mm in an adult (Epker), 11 mm in a man, 10 mm in a woman (Turvey). It grows from 9 to 12.4 mm between the age of one year and 18 years in a male and from 9 to 11.1 between the age of one and 18 years in a female (Burstone).

This thickening due to growth is less important than that which occurs at the maxillary level.

The surface of the soft tissues is crossed by a depression, a sill, and an undulation which can be explained anatomically and somewhat modified by mobilization of the skin and muscle of the labiomental region.

One can note:

a. The labiomental crease, transverse, arching, with an inferior concavity separating the lip from the chin. This corresponds with the upper portion of the mental eminence and the vestibular cul-de-sac. The median, criss-crossed fibres of the muscle of the crest, interdigitate with the obicularis of the lip above, and the fibrous tissue cushion of the chin below which is a small space occupied by a constant adipose accumulation (Magnier). Many have tried to define the origin of this crease which has been called the 'sublabial contraction' by Ricketts, with particular emphasis on the muscular insertion that is sometimes a true independent horizontal muscular cord (Gray's anatomy, Hamula).

The arguments in favour of this muscular origin are supported by postsurgical observations (the vestibuloplasty of Hamula).

b. The central depression takes the form of a vertical crease or dimple. This is caused by the cutaneous insertion of the crest ligament and also certainly of central fibres of the muscle of the crest. The ligament, a true tendinous intermuscular raphé arising from the mental symphysis, has superior fibres which arrive at the labial mental crease; the inferior fibres form the spans which contain the adipose accumulation. The form and depth of this projection are functions of the fibrous cord and the adipose cushions. A familial character is often recognizable.

c. A premental bursa near the point of the chin has been described by Richet, a serous bursa between the periosteum and the muscular layer (the layer formed by the muscle and ligament of the crest in its inferior portion and several fibres of the platysma).

d. The jugo-submental crease. The independent transverse muscle of the chin, or the emanation of the triangularis muscle of the lips, is constant in two-thirds of cases, and more often in the female, and it forms by its lower border a posterior extension of the anteroinferior angle of the triangularis to the median subhyoid region, and the submental crease. This crease extends over the surface and separates the cheek from the chin. It appears in the embryo at 10 weeks, coming from aberrant fascicles of the triangularis (Futurama) and Sappey looked upon it as a tensor of the skin of the mental region.

3. Morphology of the chin — the interrelationship between bone and soft tissue.

Except for the phenomenon of ageing, it is difficult to appreciate anomalies of the chin when one looks for those that are due particularly to abnormalities located at the soft tissue level, without trying to understand the interrelations and eventual correspondence of the bone and soft tissues. These interrelationships, which have an effect on the eventual morphology of the chin, can be often quite variable as is demonstrated by a number of observations which often conflict:

— As was shown by Subtelny, during growth the soft tissue of the labiomental region (as opposed to that of the maxillary region) is directly linked to osseous modifications; thus the bony chin and the tegument develop in parallel fashion associated with anterior mandibular growth and dental eruption (Figs. 6. and 6.2).

— On the other hand, further examples show that the soft tissues neither accompany nor attach directly to the skeletal profile.

a. In dentofacial orthopaedics the muscular and skin covering can either aggravate or mask an anomaly of the skeletal profile (Aloe).

An example could be provided by comparison of the two Burstone class II, type 1s (Fig. 6.3) who show identical dentoskeletal anomalies but each with quite different contours of the soft tissue envelope, justifying the study of the variations in the distribution of the soft tissues as a function of the 'ideal mean' as this same author has proposed.

b. In addition, many studies discuss the displacement of the soft tissues secondary to osseous displacements resulting from osteotomies and show the difficulties in appreciating the redraping of the soft tissues.

The soft tissues react in a different manner according to the amplitude and the direction of

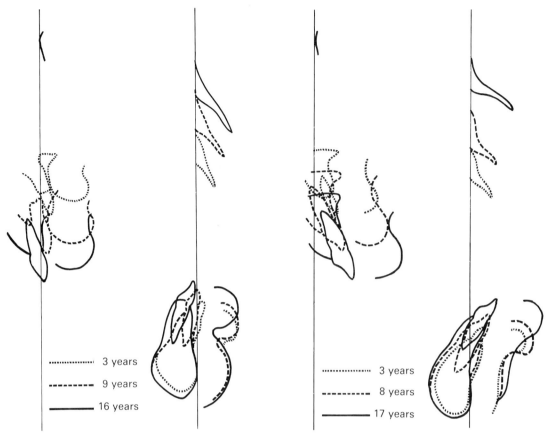

Fig. 6.1 Superimposed tracings of the same patient at different ages showing the evolution of the central incisor and the alveolar process in relation to the facial plane with the labial position which developed.

Fig. 6.2 Superimposed tracings of the same patient at different ages showing the dental alveolar protrusion in relation to the facial plane with the labial protrusion correlated (From Subtelny & Rochester 1959).

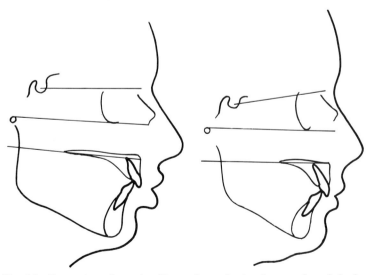

Fig. 6.3 Comparison of two class II, type I cases having the same dentoskeletal anomalies but different cutaneous profiles (From Burstone 1958).

advancement or recession, even for an osteotomy of which the outlines are identical, e.g. Le Fort I, McCarthy; Lines, advancing or setting back the mandible; modification of the lower lip.

The elasticity of different regions is variable (Hershey-Smith: the operation of mandibular set-back with a ratio of 1 for bone and 0.7 for the soft tissues of the chin; 0.6 for the lower lip and 0.2 for the upper lip).

The study is often done in the form of a simple ratio which involves only one anatomical region taken in its totality and does not take account of specific movements of each portion or isolated point. The movements of the soft tissues are in effect subtle and complex (lengthening, compression, shearing) and exist even if the osseous displacement is apparently simple, like an anteroposterior maxillary or mandibular movement. With more complex computer-assisted studies done with multifunctional tensor analysis of elasticity perhaps one will be better able to appreciate these modifications even at a distance from the site of the osteotomy. Modifications affecting the chin have been observed after osteotomies at the level of the maxilla, as in the case of the long face syndrome (Merville) where there is a vertical maxillary excess or a hypsomaxillar anomaly and an exaggerated opening of the gonial angle (amblygonia). The chin which seems aesthetically to lack good contour preoperatively can, after a procedure carried out at the maxilla (Le Fort I elevation and advancement), show a mental protrusion with the creation of a true labiomental fold (and is explained, in large part, by the notion of autorotation of the mandible).

On the other hand, with significant advancement genioplasties, in particular with the 'jumping' of the basilar fragment onto the upper fragment with creating a significant displacement, the labiomental fold is only mildly modified and usually in a haphazard fashion (Scheiderman, Gallagher-Bell, Tulasne, Depreaux).

The interrelationships between the bony chin and the soft tissue chin are, therefore, very variable and have a multifactorial origin. In particular the following have an effect:

(i)   The neuromuscular adaptation to a new maxillomandibular equilibrium.

(ii) The mandibular movements after a procedure on the maxilla (autorotation);
(iii) The notion of a cutaneous memory in the mandibular region;
(iv) Individual elasticity;
(v)   Cervical posture (cervicomental angle and position of the hyoid bone);
(vi) Particularly the position and the modifications in the position of the teeth (overbite, overjet);
(vii) The position and the modifications in the position of the lips (certainly the lower lip but also the upper lip, due to labial interferences): the notion of pseudocompensation of a dentoskeletal anomaly (Aloe–Burstone) and hence the necessity of analysing the chin in the complete rest position of the lips.

These different factors take account of the difficulty in analysing an anomaly of the soft tissues of the chin.

ANOMALIES IN THE SOFT TISSUES OF THE CHIN

1. Modification of the volume of the chin

The activity of the peribuccal muscles, particularly the muscles of the crest and the orbicularis of the

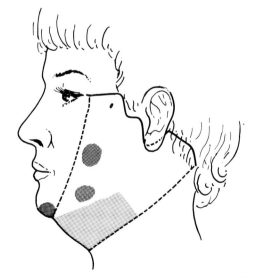

Fig. 6.4   Schematic representation of the steatomeric (fatty) facial separation determining the fat excision during rhitidectomies (From Regnault 1981).

lips, can vary depending on the dentoskeletal anomaly (Burstone, Lowe-Takada).

In particular, the hypertrophy by hyperactivity — the cause of a mental bulge — is seen in certain maxillomandibular dysmorphoses:

a. certain class II type I conditions (Wolford–Epker) where the prominence of the soft-tissue chin extends beyond the plane of the lower lip;
b. in the long face syndrome with a vertical excess of the inferior portion of the lower part of the face (hypsogenia) (McBride–Bell), a vertical excess of the maxilla (Double bump–Kawamoto) or an anterior incisive (openbite);
c. in bimaxillary protrusion.

On the other hand, the appearance of the chin remains normal in bilateral facial paralysis in the case of the Moebius syndrome (Couly).

Finally, hypertrophy can be caused by the extension of a tumour (notably bony into the soft tissues) or due to a tumour located in the muscles, e.g. a haemangioma in the muscle of the crest — Ingalls.

2. Modification of the shape of the chin

a. Modification of the labiomental fold.

Following the classification of Hamula, several aetiologies can exist in isolation or association.
  • With underlying osseous anomalies:
— the soft tissue chin can be flattened with a labiomental fold which is poorly defined, and a bony chin, which is often flat as well as high or long, with a lysis of vestibular bone, and a muscular flattening seen in the long face syndrome associated with hypsogenia (Tulasne, Roulo, Delaire, Plenier, McBride–Bell);
— or the chin may have a very deep labiomental fold in the class II conditions (type I, Wolford–Epker), and in the short face syndrome (type I, Freihofer) in which case the lower portion has the aspect somewhat of a rocking chair.
  • With dental anomalies (overjet, alveolar protrusion etc), one should underline the aetiologies due to the soft tissues themselves:
  Extrabuccal

— labial incompetence or separation;

— thick lips, labial aversion;
— hypertrophy of the soft tissues of the chin, particularly muscular hypertrophy.

  Endobuccal

— depth of the vestibule;
— height of the gingival attachment;
— position of the mucosal frenulum.

In the correction of dentoskeletal anomalies, some authors have described surgical techniques to modify the depth of the crease (to diminish it in most instances in cases of endobuccal anomaly and muscular attachment: achieved by advancement and deepening vestibuloplasty, sometimes with excision of the muscle of the crest (Hamula).

Changes in contour after vestibuloplasty have been studied elsewhere (vestibuloplasty requiring skin or mucosal grafts) (Adaway, Hillerup, Turvey–Epker), and among other things, a better projection of the chin, an eversion of the vermillion and particularly an attenuation or disappearance of the excessively pronounced labiomental crease have been noted.

b. Modifications of the submental crease and the cervicomental angle.

These are studied in the Chapter on Ageing although some of these modifications can be seen in young subjects with characteristic familial traits such as:

(i) ptosis of the chin, encountered even without any associated dentoskeletal anomaly, e.g. mandibular protrusion, and thus caused by a localized excess of soft tissue on the point of the chin;
(ii) the heaping up of tissue of the cervicomental region seen in young subjects with a skeletal anomaly (mandibular protrusion, anterior position of the hyoid and of the superhyoid muscles — Marino, Turvey–Epker, Collins).

c. Modifications of the midline crease.

This is often not present and there are requests for its creation for aesthetic reasons. Cinelli, in particular, reported a technique of making a 'cleft chin' by the simple subcutaneous excision of tissue, preferable according to him to a skin incision (excision) or to bony remodelling.

3. The chin and ageing

a. With ageing there is:

(i)   a muscular relaxation (in particular of the platysma which forms a cord in the cervicomental angle and which can decussate across the midline;
(ii) an acquired ligamental laxity;
(iii) a localized fatty accumulation;
(iv) the creation of folds and sagging in the mental and submental areas.

Other factors play a role:

(i)   retrogenia;
(ii) a position of the hyoid bone which is too low and anterior;
(iii) the edentulous state, associated with the loss of alveolodental support of the lip;
(iv) mandibular autorotation associated with the edentulous state.

b. Different clinical forms

Different clinical forms are encountered with different possible classifications. Schematically, one can see in varying degrees:

(i)   Ptosis of the chin or 'Witch's chin' with a falling of the point of the chin and an excess adiposity at this level, separated from the submental region by a very pronounced submental crease;
(ii) A double chin with a localized accumulation in the cervicomental angle ('double chin'), the 'Turkey-Gobbler' deformity where there is a filling of the cervicomental angle (normal: 105 to 120° by excess skin, subcutaneous fat and platysma muscle, with sometimes an anomaly in the position of the hyoid bone.

c. Different criteria

Different criteria have been proposed to appreciate the excess skin, fat, and state of the platysma, as well as the position of the hyoid bone and the bony chin (micro- or retrogenia), and lead to isolated procedures or associated corrections of the anomaly such as:

(i)   Skin excision, discussed by numerous authors (Millard, Courtiss for whom the excess is only apparent), and calling for different types of plastic procedure (the inferior pedicle of Rees, the Z plasty, W plasty, and T–Z plasty of Cronin, and the plasty in the shape of an S (Morel-Fatio) which in most cases leaves a large scar;
(ii) Excision of fat, either by liposuction (Illouz, Teimourian, Courtiss) — an easy method which is noninvasive and is used irrespective of age and degree of cutaneous elasticity (Courtiss), or by lipectomy with a submental incision permitting the excision of subcutaneous fat, and sometimes subplatysmal and submandibular fat;
(iii) Plication of the platysma:
      — laterally by suture of the posterior border to the sternocleidomastoid facia
      — anteriorly by suture of the platysma in the midline from the symphysis to the thyroid cartilage (in particular for types I and III of Cardoso), or by associated excision: anterior myectomies, total or partial myotomies with the creation of different types of flaps or muscular 'hammocks'
(iv) Correction of the retrogenia, either by placing an implant (Ellenbogen, Courtiss) using a submental incision or by horizontal basilar advancement osteotomy through an intraoral approach (Wolfe);
(v)   Reposition of the hyoid bone in a higher and more anterior position to optimize the results of the genioplasty and close the cervicomental angle (Collins-Epker).

IN CONCLUSION

Analysis of the morphology of the chin and of the submental region requires not only the study of the different components, (the skeleton and the soft tissues) but also adjacent anatomical structures (the lip and the neck) and structures at a distance (the relationship between the nose and chin).

In this manner, one can best establish the indications and the associated procedures which should be carried out on the soft tissues or on the bone in the framework of a global orthomorphical surgery of the entire face. This should lead to procedures of the soft tissue, complementary maxillofacial orthopaedic procedures (labiomental fold) or, on the other hand, to bony procedures associated with aesthetic surgery of the soft tissues (advancement genioplasty).

BIBLIOGRAPHY

Burstone C J 1958 The integumental profile. American Journal of Orthodontics 44: 1

Regnault P 1981 Facial and Submandibular lipectomy with upward rhitidectomy. Journal of Maxillofacial Surgery 9: 247–252

Subtelny J D, Rochester N Y 1959 A longtudinal study of soft tissue facial structures and their profile characteristics defined in relation to underlying skeletal structures. American Journal of Orthodontics 45(7): 481–507

7. Vascularization of the bony chin

B. Ricbourg

INTRODUCTION

When Ruysch, in 1701, for the first time injected the terminal watershed of the lingual artery, it is unlikely that he could have imagined the possibilities of contemporary maxillofacial surgery.

Two hundred and eighty-nine years later, our knowledge of quantitative arterial vascularization of the bony chin is still poor.

The arterial architecture has been considerably studied, however, and we will discuss several aspects of this.

THE ARTERIAL SUPPLY

1. The arteries in question

These are all branches of the external carotid system.

a. The inferior dental artery (Fig. 7.2)

— arises from the trunk of the internal maxillary;
— penetrates the mandible at the level of the spine of Spix;
— passes through the dental canal;
— divides into two branches at the level of the premolars: a mental branch which comes through the mental foramen and supplies the integuments of the chin, and an interosseous incisive branch for the alveolar bone and the incisive–canine segment.

b. The sublingual artery (Figs. 7.1 and 7.3–7.7)

— arises from the bifurcation of the lingual artery at the level of the anterior border of the hyoglossus muscle;
— proceeds anteriorly in a muscular arcade

located outside the canal of Wharton, and passes under the sublingual gland;
— terminates in the fibres of the genioglossus muscle near its osseous insertion, as well as in the overlying gingival mucosa.

c. The ranular artery (Figs. 7.1, 7.2, 7.6 and 7.7)

— small branches come out of the ranular arch and are distributed to the retro-alveolar mucosa of the parasymphyseal regions.

d. The facial artery (and its collateral branches) (Figs. 7.2 and 7.5)

— arises directly from the external carotids;
— gives multiple collateral branches all along its course. For the chin, it arrives by the inferior coronary branch (of which one portion courses towards the integument of the chin), and the submental artery (which

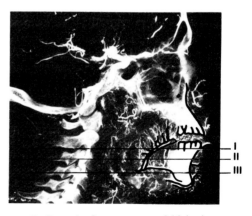

Fig. 7.1 Radiograph after opaque arterial injection; showing a vertical section of a head injected densely in the paramedian area and the three levels of horizontal section in (Figs. 7.2–7.4).

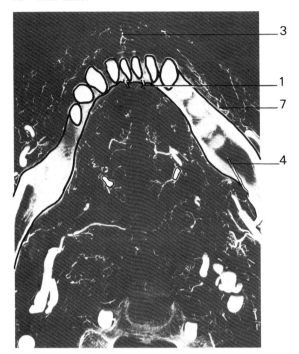

Fig. 7.2 Radiograph after opaque arterial injection. Horizontal section at level I (Fig. 7.1). 1. ranular artery; 3. collateral facial branch; 4. inferior dental artery; 7. mental artery.

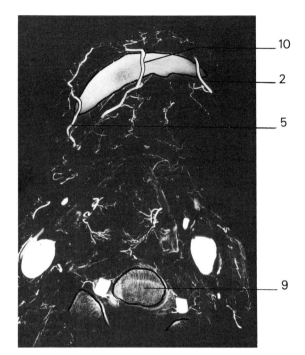

Fig. 7.4 Radiograph after opaque arterial injection. Horizontal section at level III (Fig. 7.1). 2. sublingual artery; 5. submental artery; 9. odontoid apophysis; 10. anastomosis 2–3.

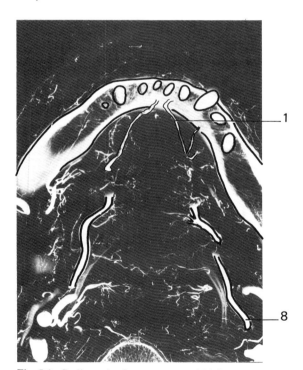

Fig. 7.3 Radiograph after opaque arterial injection. Horizontal section at level II (Fig. 7.1). 1. sublingual artery; 8. lingual artery.

follows the mandibular border and terminates at the level of the parasymphyseal region by anastamosing with the branches from the sublingual).

2. The arterial networks

a. The medullary network (Fig. 7.5)

— is located in the diploe or the cancellous bone of the mandible;
— is supplied by the inferior dental artery in a small part and by the sublingual artery in a greater part;
— is seen in small passages destined for the dental roots and also forms small blood lakes in the alveolar bone.

b. The gingivoperiosteal network

— corresponds to the 'attached' zone of gingiva;
— the gingiva and the periosteum are intimately connected;
— is supplied by the sublingual and ranular arteries behind, and the facial artery and

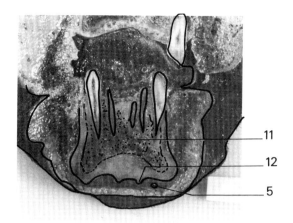

Fig. 7.5 Frontal section of the bony chin. 5. submental artery; 11. alveolar bone; 12. inferior genial apophysis.

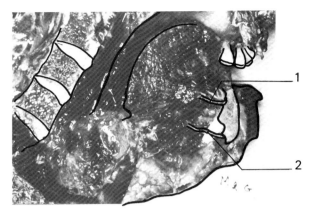

Fig. 7.6 Dissection after arterial injection. Median sagittal cut, with 'intra-oral' view. 1. branch of the ranular artery; 2. branch of the sublingual artery.

mental branch of the inferior dental artery in front;
— forms a small, interlinked network covering the entire 'uncovered' region of the symphysis.

c. The muscular periosteal network

— corresponds to the 'covered' region of the mandible;
— is made up of an entire series of branches which are initially muscular and penetrate the bone perpendicularly, and other branches which form a network included in the periosteum;
— is supplied by the facial collaterals in front and the sublingual artery behind.

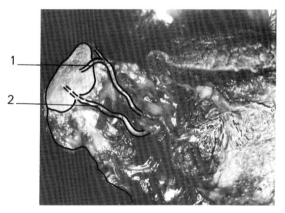

Fig. 7.7 Dissection after arterial injection. Median sagittal cut, internal view. 1. branch of the ranular artery; 2. branch of the sublingual artery.

3. Anatomical points of surgical interest
(Figs 7.6 and 7.7)

It is evident that the architecture that has been described is theoretical.
Anastomoses exist:

— between the different networks;
— from one side to the other;
— are regulated by the 'law of balance' which states that an insufficiency of one arterial branch calls for a compensatory hypertrophy of a neighbouring branch.

A point which should be emphasized, since it is not well described in the classical anatomy books,

is the considerable importance of the terminal branches of the *sublingual arteries*. They penetrate the bone in many areas through two vascular passages: one located between the two upper genial apophyses, the other median between the two inferior genial apophyses, which is to say between 6 to 8 mm from the lower border of the mandible. The superior branch has a calibre greater than 1 mm and its supply is thus quantitatively important.

There exist only minuscule vascular holes on the anterior surface the symphyseal and parasymphyseal regions.

The terminal branches of the inferior dental artery are very frail and have an exclusive dental

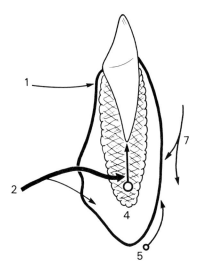

Fig. 7.8 Parasymphyseal section (schematic). 1. Ranular arcade; 2. sublingual; 4.inferior dental; 5. submental; 7. mental branches.

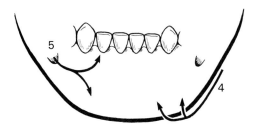

Fig. 7.9 Anterior surface of the mandible (schematic). 4. submental artery; 5. mental branches.

Table 7.1

Medulary network	→ Sublingual artery +++
	→ Inferior dental artery
Gingivoperiosteal network	→ Sublingual artery
	→ Ranular arcade
	→ Mental branches
Musculoperiosteal network	→ Sublingual artery +++
	→ Branches of the facial artery

destination. The alveolar bone is supplied by the medullary network.

To sum up (Figs 7.8 and 7.9, table 7.1):

1. The vascular passages are schematically:
   a. large vessels on the posterior surface
   b. weak vessels and by a periosteal network on the anterior surface
2. A mandibular dissection should be done only on the anterior surface.
3. Surgical osteotomies of the bone which involve the least amount of devascularization are carried out on a horizontal plane between 9 and 11 mm from the lower border of the symphysis.
4. Sections of the alveolar bone lead only to dental denervation, and the bone consolidates rapidly.

VENOUS DRAINAGE

1. The three networks of blood collection follow the arterial system:

   — medullary
   — gingivoperiosteal
   — musculoperiosteal

2. Venous drainage:

   a. along the posterior surface, the vessels form and come together and follow the arterial axis towards the lingual and ranular veins then the thyrolinguofacial trunk and from there progress towards the internal jugular vein.

   b. on the anterior surface there are two possibilities: drainage occurs either directly into the anterior jugular veins, or by anastomoses with the posterior network and from there towards the interior jugular vein.

LYMPHATIC DRAINAGE

This occurs in two ways:

1. to the submental nodes in the anterior jugular chain;
2. to the submaxillary nodes in the internal jugular chain.

BIBLIOGRAPHY

Couly G Anatomie maxillo-faciale. Prélat, Paris.
   pp 81–86 and 87–96
Poirier P, Charpy A 1899 Traité d'anatomie humaine. Tome
   I, Ostéologie, Masson, Paris. pp 520–527
Poirier P, Charpy A 1901 Traité d'anatomie humaine. Tome
   IV, Tube digestif, Masson, Paris. pp 124–126
Poirier P, Charpy A 1902 Traité d'anatomie humaine. Tome
   II, Angéiologie, Masson, Paris. pp 661–679 and 685–686

Ricbourg B 1974 Artères et veines cutanées de la Face et du
   cuir chevelu. Thèse Doctorat en Médecine, Paris.
Ricbourg B 1985 Quelques aspects de la vascularisation du
   menton osseux. Com. XIIe Congrès de A.F.C.M.F.,
   Nancy, 25–27 April 1985
Ricbourg B et al 1979 A propos de certains aspects de la
   vascularisation de l'orifice buccal. Annalex de Médecine
   de Nancy 4: 1213–1220

8. Minor and major deviations of the chin — possibilities of treatment and problems encountered

H. P. Freihofer

INTRODUCTION

Every extreme deviation of the chin, whatever its direction in space, is poorly accepted by the patient, and one can understand why there is demand for correction. It is different when the deviation is more discrete; in those cases where the asymmetry of the face is even noticed by the patient, they rarely complain about it, in contrast to those patients with a pronounced laterognathia. In discrete deformities, the shape of the chin, its height and the configuration of the labiogenial fold seem to be more important to the patient.

In asymmetries, preoperative studies involve clinical examination of the face and analysis of a lateral cephalometric film. An enlarged photograph of the patient, (McBride & Bell 1981) we feel, is not essential.

The operative technique of genioplasty which we employ is practically always that in which the genial muscles are not disinserted which avoids the exaggerated resorption of the bony segment.

In most cases, sliding osteotomies are used for the correction of asymmetries (Obwegeser 1957, 1958; Koele 1961).

We shall now look at the various possibilities. The discussion of results will concentrate on the aesthetic aspects.

THE SIMPLE GENIOPLASTY

A genioplasty here aims at the total correction of the deformity. The indication is not always as easy as one might think. All too often, the mental deviation is combined with an asymmetry of the mandible, which can often be quite discrete itself, and in this situation one should decide on a rotation of the mandible in its entirety rather than a simple genioplasty (Fig. 8.1).

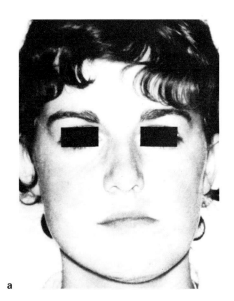

a

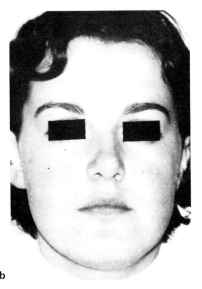

b

Fig. 8.1 **a.** Asymmetrical prognathism. **b.** Very good symmetry achieved by mandibular rotation.

In osteotomies with translations of the chin, it is always recommended to preserve as many of the muscular attachments and soft tissues on the lower border of the mandible as possible, including the platysma muscle. Adhering to this advice, one can be more certain that the soft tissues will follow (Fig. 8.2).

If the posterior border is completely detached and the segment only has a posterior muscular pedicle, the overcorrection necessary can be considerable (Fig. 8.3).

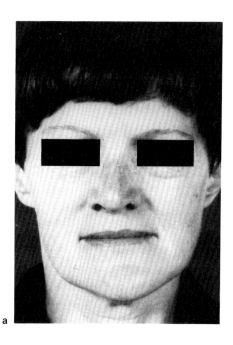

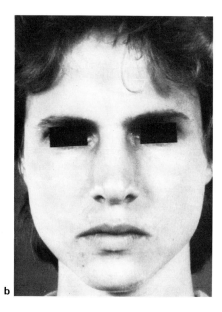

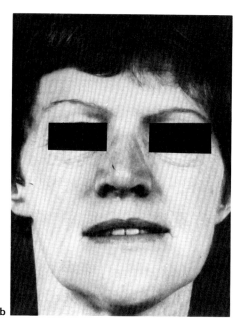

Fig. 8.2 a. Asymmetry of the mandible with good occlusion. **b.** Asymmetrical osteotomy and translation of the chin with minimal overcorrection. Good symmetry.

Fig. 8.3 a. Asymmetry in an edentulous patient. **b.** Insufficient translation not giving the best possible result.

THE GENIOPLASTY ACCOMPANYING OTHER OPERATIVE PROCEDURES

We mean by this movements of the chin going in the same direction as those of the mandible or the upper jaw. If the deviation of the chin is caused by mandibular asymmetry it is normal to consider a rotation of the mandible, be it prognathic and large or, on the contrary, deficient. This approach makes it possible to correct an asymmetry of the lower portion of the face more satisfactorily. In many cases, however, the occlusion can cause problems, and thus one can observe with this type of operation a clear difference between the correction of the occlusion which is felt to be satisfactory and the correction of the deviation of the chin which has not followed quite as well because of the general form of the mandible.

One needs, therefore, to return for further alterations of the osseous base, and the genioplasty here becomes of paramount importance (Fig. 8.4).

We should note here that preoperative planning of the necessary procedures is more difficult to carry out in such cases than in simple transpositions of the chin, because here we have to take into account the movement of the mandible itself. The correction of the malocclusion of the mandible is relatively easy to predict, but it is more difficult to predict the eventual position of the point of the chin, particularly if tilting movements are necessary. In fact, the additional movement of correction of the chin cannot be predicted with precision during the preoperative study.

It should be emphasized that rotational movements of the mandible can cause asymmetries of the soft tissues in the region of the mandibular angle which are not corrected by a genioplasty. On the other hand, vertical osseous asymmetries, such as one encounters in hyperplasia of the condyle and the mandible, sometimes require additional osteotomies of the horizontal ramus of the mandible extending back to the mandibular angle (Fig. 8.5).

In severe asymmetries with a deficiency of bone, individual solutions need to be found for each particular case. These may include maxillary osteotomies, osteotomies of the mandible and chin, and also cartilage grafts of the mandibular border (Fig. 8.6).

The asymmetries in craniofacial microsomias

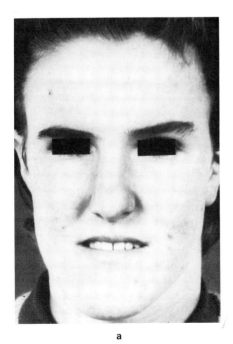 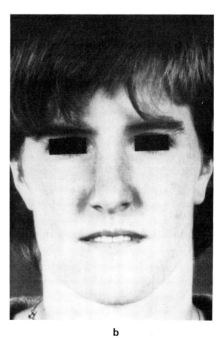

a   b

Fig. 8.4 **a.** Asymmetrical prognathism. **b.** Mandibular rotation and complimentary translational genioplasty and asymmetrical osteotomy performed at one session resulting in satisfactory symmetry.

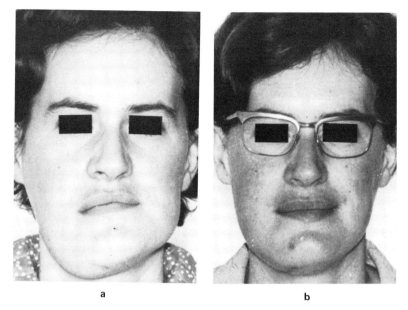

Fig. 8.5 **a.** Hyperplasia of the mandible. **b.** In spite of a rotation of the mandible and osteotomy along the horizontal ramus, symmetry has not been achieved in one operative session. A further modelling procedure will be necessary.

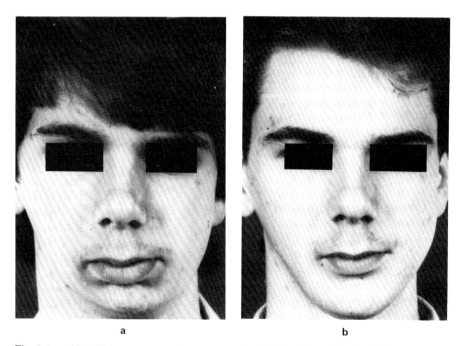

Fig. 8.6 **a.** Mandibular asymmetry after trauma and ankylosis. **b.** The situation did not permit a rotational osteotomy of the mandible and the result, which was not completely satisfactory, was obtained by extensive rib grafting and a translation tilting of the chin with an interpositional bone graft.

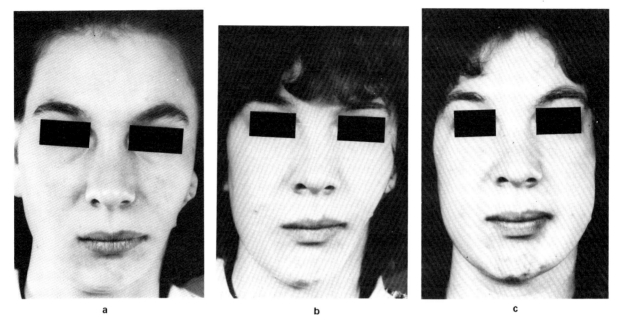

Fig. 8.7 a. Craniofacial microsomia. **b.** After achieving symmetry of the skeleton and alignment of the mid points by rotation and tilting of the maxilla, the mandible and the chin (Obwegeser 1974), the deficit of the soft tissues is accentuated. **c.** Final correction by free fat transplantation.

present additional problems at the level of the chin. First, the chin segment itself may have an asymmetrical form, where the vertical asymmetry of the two sides can be compensated for either by an asymmetrical osteotomy or the asymmetrical interposition of bone, or if the height of the chin permits, by a 'propeller' genioplasty (Sailer 1983). Equilibrium and symmetry in the horizontal plane are difficult to obtain. A crescent-shaped segmentation of the chin, as proposed by Obwegeser (1958), can provide the solution. One must bear in mind also that since there is both a deficiency of the skeleton and of the soft tissues, providing skeletal symmetry may make the deficit of the soft tissues even more apparent. Correction of this then becomes necessary. This can be done by pedicled muscular flaps, anastomosed flaps or by free fat grafts (Fig. 8.7).

THE GENIOPLASTY AS A COMPROMISE

All types of compromise genioplasties can be imagined. Take as an example the case of a patient having mandibular asymmetry but who, for whatever reason — for example, a normal occlusion — is not searching for anything but the correction of the asymmetry of the chin.

It is also possible that a genioplasty would be indicated to provide a temporary amelioration for a problem in evolution. The proper moment for definitive correction not having yet arrived, one can in the mean time improve the deformity with a genioplasty. In this way, one helps the patient with a small procedure that is even reversible if analysis at the time of the definitive correction demands it (Fig. 8.8).

THE HARMONIZING GENIOPLASTY

This is done in cases where the movement of the correction goes in the opposite direction to the movement of the mandible to obtain a perfect result. It is not often necessary in asymmetries, but finds its application more frequently in anteroposterior disharmonies where it is the accompaniment, for example, of a maxillary advancement or a chin advancement. Nevertheless, one must keep in mind this possibility in asymmetries as well, since the patient will not accept having an asymmetry on one side corrected

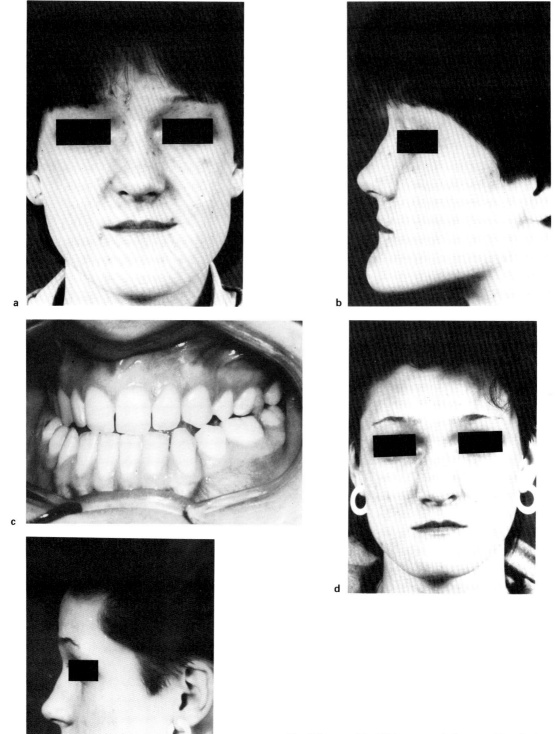

Fig. 8.8 **a.** and **b.** Mild asymmetrical prognathism during evolution in a girl 16 years of age. A definitive correction is not advised at the moment, but given the psychological problems an amelioration is desirable. **d** and **e**. Compromise genioplasty which completely satisfied the patient. The eventual final growth of the mandible is still expected to occur.

by the creation of an asymmetry on the other side, even if it is less pronounced than the original deformity (Fig. 8.9).

DISCUSSION

For several reasons, the correction of a deviation, either of the chin or the entire mandible is not easy to accomplish. Without entering into vertical and anteroposterior problems which can be present, an asymmetry alone poses several problems. It is evident that a definition of the mid-line of the face before the operation is of major importance. It is only with this that one can establish a good preoperative plan, and guarantee perfect execution of the operative procedure.

However, the facial mid-line can be difficult to define because of the following effect: on the occlusal plane, the movement of the mandibular rotation can be calculated in millimetres. This can-

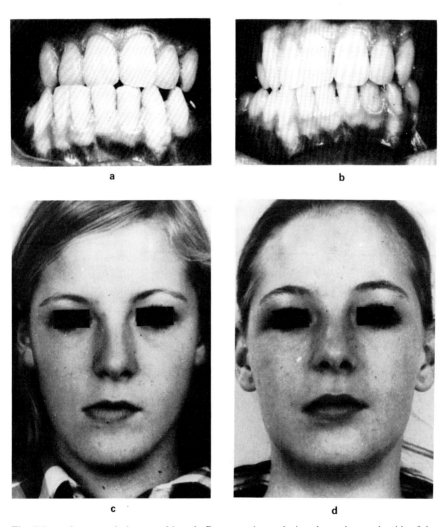

Fig. 8.9 **a.** Asymmetrical prognathism. **b.** Postoperative occlusion shows that on the side of the mandibular translation a rotation was necessary. This displaced the right side inferiorly (note the axis of the teeth). **c.** Preoperative aspect with a deviation of the chin towards the right. **d.** After the osteotomy a slight asymmetry towards the left can be noted. The rotation was greater at the level of the chin than at the level of the teeth. A harmonizing genioplasty could have compensated for this.

not be said of the movement that the chin must undergo at the same time. Particularly if tilting movements of the mandible are necessary, the operative plan and the preoperative calculations are very difficult to carry out. The translation of the chin can differ considerably from that of the mandible when it is brought into occlusion. It is for this reason that one cannot always establish before the operation if a genioplasty will be necessary or not.

The second obstacle is the following: how will the soft tissues, in the case of an osteotomy of the chin, follow the movement of the osseous segment? It has been noted that the more the soft tissues are attached to the lower border segment, the better they follow. Nevertheless, there is not an exact parallel movement between the soft tissues and bone. In these circumstances, it is better to over-correct a little.

In cases where the soft tissues are attached more or less completely, the overcorrection becomes considerable. These facts bring one to the conclusion that the result cannot be perceived until the time of the operation. This is not as easy as one might think, even if the mid-line of the face has been well-established before the operation. The nasal intubation, intraoperative oedema, and the position of the patient are factors which can considerably interfere with our interoperative appreciation.

Other problems are added to this as well. An unfavourable position of the ends of the lower border segment can be camouflaged during the operation by an undermining of the soft tissues and the intraoperative oedema. In the long term, this can cause asymmetry of the horizontal ramus, the point of the chin itself being in the correct position or where one wants it.

Major rotations can cause asymmetries of the cheek. The soft tissues are tight, and they appear swollen. This is very evident in the correction of otomandibular dysostoses which have a true deficiency of soft tissue. But even in a mandibular asymmetry which is supposedly a simple one, the same phenomenon can be seen. For all these reasons, one can say that it is impossible to guarantee in any particular case that there will be a perfect correction of the deviation of the chin in one operative procedure.

In addition, one must appreciate the possibility of a relapse of maxillary and mandibular osteotomies as well as bony resorption of the chin.

Since the genioplasty is a purely aesthetic correction, it is of great importance that the analysis is well done from the very beginning.

Since it is sometimes impossible, as we have seen, to predict everything with exactitude, it is important to explain this as well as possible to the patient and have the patient appreciate the problem. This makes it possible for us, during the operation, to make any decisions which need to be made. The preoperative evaluation has, in any event, given us direction.

Even though it is difficult to arrive at a perfect result, one should try to correct everything in one operative session. Apart from rare exceptions, we have not had patients rehospitalized because it was deemed necessary afterwards to perform an additional genioplasty. The patient and the surgeon risk having to be content with a mediocre aesthetic result if they miss their opportunity at the first, and often only, operation.

On the other hand, I have never seen a supplementary genioplasty which has aggravated the situation. However, I have seen a number which did not sufficiently correct the primary deformity.

CONCLUSION

Deviations of the chin are difficult to correct, whether they are minor or major. There are few other skeletal malformations that one encounters where the clinical examination, above all other types of analysis and calculation, plays such an important role.

The majority of cases need to be treated by a mandibular osteotomy or even by maxillary and mandibular osteotomies together. But a genioplasty combined with other osteotomies is much more often indicated than one might think in general.

The genioplasty is perhaps a detail. But when one is dealing with aesthetic problems, as everyone knows, a detail is much more than a detail.

BIBLIOGRAPHY

Koele H 1961 Korrekturen am Kinn und Kierferwinkel. In: Schuchardt K (ed) Fortschritte Kiefergesichtschirurgie. Vol. 7, Thieme, Stuttgart, p 165

McBride K L, Bell W H 1980 Chin surgery. In: Bell H W et al (eds) Surgical correction of dentofacial deformities. Saunders, Philadelphia, p 1210

Obwegeser H L 1957 The surgical correction of mandibular prognathism and retrognathism with consideration of genioplasty. Oral Surgery, Oral Medicine, Oral Pathology 10: 677

Obwegeser H L 1958 Die Kinnvergrösserung. Oesterreichische Zeitschrift für Stomatologie 55: 535

Obwegeser H L 1974 Correction of the skeletal anomalies of oto-mandibular dysostosis. Journal of Maxillofacial Surgery 2: 73

Sailer H F 1983 Die Propellerkinnplastik. Présenté au Congrès de l'I.A.O.S., Berlin, 1983.

9. The chin in the long face. Methods of correction

L. C. Merville

For the term 'long face syndrome', which is certainly convenient, and taken from the American vocabulary, it would seem more exact to substitute the term 'high face' taking account of the usual erect posture of the human.

Besides, in a recent article in *Plastic and Reconstructive Surgery* (Farkas et al 1985) on the vertical proportions of the face, Farkas, one of the members of the Craniomaxillofacial surgical team in Toronto, directed by Ian Munro, indeed utilized the term 'height', and not 'length' as do certain of his 'oral surgical' colleagues to describe this type of anomaly.

Nevertheless, to avoid any linguistic or semantic quarrel, it seems perhaps better to return to a Greek etymology which can be accepted in all languages having this cultural origin. To designate a facial dysmorphia characterized by an excess height, the term of hypsoprosopia (from the Greek *hypsos* = height and *prosopos* = face) is proposed, which is analogous with the term hypsocephaly, already found in medical dictionaries to describe an excess height of the cranial vault.

In the hypsoprosopia, the chin is always located too low, and thus participates directly in the expression of the vertical excess of the face of which it constitutes the lower contour.

The excessively low position of the chin can be the result of several things. It can be a consequence of an anomaly of form in the mental region itself where the vertical axis is excessively developed in height thereby constituting a hypsogenia. In this case there is a true dysmorphia of the chin.

On the other hand, the chin can be located in an excessively low position and still, at the same time, have a satisfactory skeletal form. This is the effect of an anomaly located in another sector of the maxillomandibular complex, such as an anterior interincisive vertical open bite, or an excess height of the maxilla (hypsomaxillar anomaly). Here there is not a true dysmorphia, but rather a vertical dystopia of the chin.

As a consequence, the methods of correction are quite varied, depending directly upon the anatomical diagnosis of the causative lesion. The correction of a deformity lowering the chin thus most often requires a direct osteotomy of the chin, in general segmental for a dysmorphia of the chin, whereas an osteotomy needs to be carried out at a distance in the case of a genial dystopia, recognizing at the same time, of course, that a dysmorphia and a dystopia can both be present together.

DYSMORPHIAS OF THE CHIN

The most simple form of dysmorphia of the chin, creating a hypsoprosopia by lowering the chin, is hypsogenia. The excessive vertical dimension of the chin has quite a deforming effect on the overall facial morphology.

This anomaly is easily corrected by a double horizontal osteotomy of the chin with a resection of the intermediate segment, the height of which one has carefully calculated beforehand. This type of procedure seems much more preferable to the simple removal of a fragment from the lower border, since it permits the conservation of all the muscular insertions at its level. Carried out of course through an intra-oral approach, it at times requires a great deal of care to preserve the mental nerves, as they often emerge very close to the superior section (Fig. 9.1).

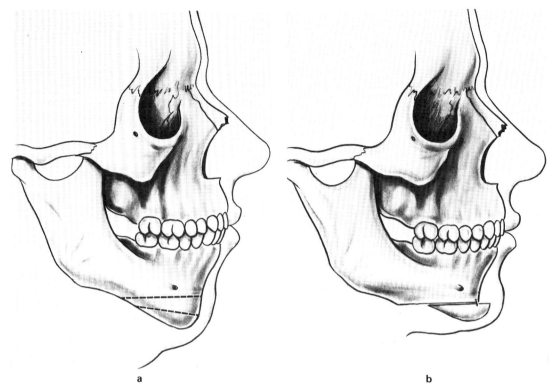

a   b

Fig. 9.1   Correction of a hypogenia by horizontal intermediate resection permitting the elevation of the basal fragment. A movement of advancement is involved here to correct simultaneously a certain degree of retrogenia.

It is rare, in fact, that hypsogenia is an isolated condition; very frequently it is found to be associated with a retrogenia (Fig. 9.2a, c). This is a typical indication for a basilar horizontal osteotomy with advancement and then elevation (Obwegeser, Converse). The inferior fragment preserves its genial muscular insertions and is placed in front of the mandibular body. The level of the osteotomy determines the extent of diminution of the vertical excess. To adapt perfectly to the convexity of the symphysis, the concavity of the basilar fragment should be burred down to the appropriate curvature. The thickness of the latter segment assures the correction of the retrogenia. If necessary, of course, it can be reduced. The best method of fixation, assuring stability and perfect contact, is provided by the use of screws (Fig. 9.2f).

Along with this excess in mental height there is frequently an associated alveolar anomaly, with an inferior displacement of the incisive alveolar seg-

ment (Fig. 9.3a) which further augments the vertical dimension of the face.

This association lends itself completely to the osteotomy proposed by Koële (1959) which combines a U-shaped alveolar osteotomy, elevating the incisive–canine segment, with a horizontal basilar osteotomy of the resection type, the inferior resected fragment being immediately used as an intermediate interpositional graft.

The operation is conducted through an endobuccal approach. Anaesthetic intubation is carried out through a nasal approach, and the alveolar fragment elevated to the proper level to provide a good dental occlusion; it is maintained in this position by use of a mandibular splint. The movement of elevation inevitably creates an inferior osseous defect. The hypsogenia is then corrected by a basilar resection. The inferior fragment is carefully preserved and immediately reintroduced into the inferior bony defect, having been appropriately tailored to its dimensions. This

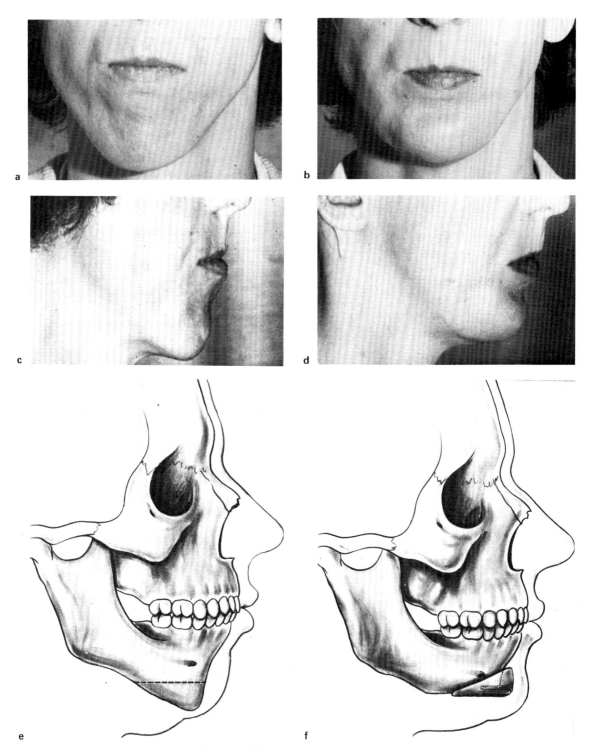

Fig. 9.2   **a.** and **c.** Hypogenia and retrogenia; **b.** and **d.** After basilar osteotomy (note also the reduction of the laterogenia); **e.** and **f.** Horizontal basilar osteotomy causing advancement and elevation, correcting at the same time the excess height and retrusion of the chin.

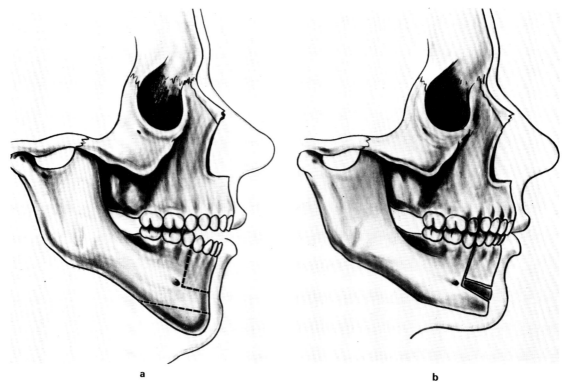

a   b

Fig. 9.3 a. Hypsogenia associated with an incisive–canine infra-alveolar anomaly. Outline of the double alveolar and basilar osteotomy of Koële. **b.** Diminution of the mental height by inferior resection of the vertical excess. Reduction of the infra-alveolar anomaly by elevation of the incisive–canine segment. Filling in of the defect created by this alveolar elevation using an interpositional bone graft taken from the resected inferior fragment.

consists, therefore, of a double osteotomy with a three-stage elevation, the lower border becoming the intermediary and vice versa (Fig. 9.3b).

To a greater degree in these dysmorphias of the chin, associated with a hypsogenia, not only is an inferior alveolar displacement found, but also a retrogenia. The combination of these three anomalies, therefore, results in an important increase in facial height, and the deforming effect on the labiocervicomental profile is even more accentuated by the posterior displacement of the chin (Fig. 9.4a, c). The pre- and postoperative drawings shown in Figure 9.4e, f illustrate the modification which we propose to the Koële procedure, leading to an arrangement which we call '3 floors with a terrace'. A third horizontal osteotomy is carried out, making possible a second harvesting of bone at the level of the resected inferior fragment to correct the hypsogenia. Placing

it in front of the body of the mandible corrects the retrogenia (Fig. 9.4b, d).

DYSTOPIA OF THE CHIN

If an increase in facial height can be caused by a dysmorphia of the chin, a lowering of the chin can also, as we have already seen, be the consequence of an anomaly located in another area of the maxillo-mandibular skeleton. This is essentially the case in anterior vertical interincisive open bites, where there is an excess height of the maxilla. The necessary elevation of the chin to correct the morphological abnormality thus depends directly on the correction of the adjacent lesion.

In the case of a vertical anterior open bite, the therapeutic plan is not always the same because there are several anomalies located in quite different areas which can be the origin of this

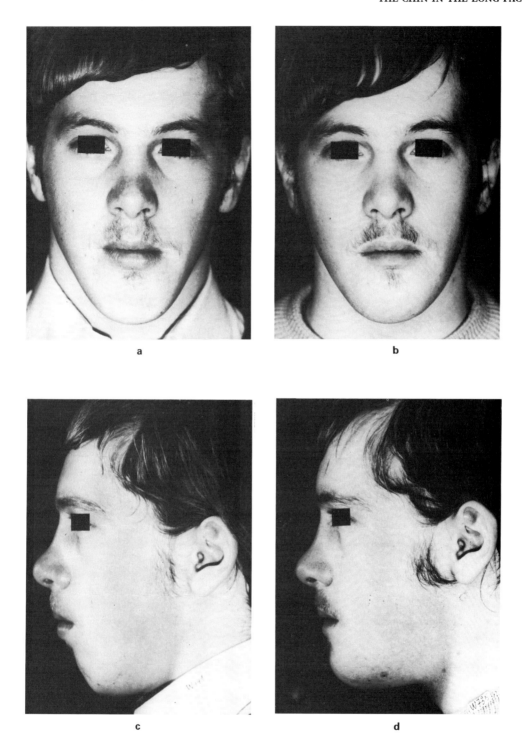

Fig. 9.4 **a.** and **c.** Considerable hypsogenia associated with a retrogenia and an inferior incisive–canine infra-alveolar anomaly, seen here in frontal and profile views; **b.** and **d.** Modifications of the morphology of the chin after a triple segmental osteotomy (note also on the frontal view a correction of a slight laterogenia).

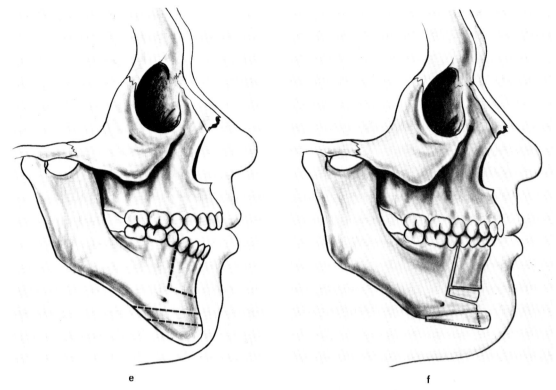

e   f

Fig. 9.4 **e.** and **f.** Schematic tracings in the profile view of this triple osteotomy, and the reduction by a '3 floors with a terrace' arrangement.

interincisive gap. Therefore, one needs to make a precise topographical diagnosis from which one can proceed to the proper osteotomy.

Three malformations tend to be responsible for such an open bite, thereby moving the chin to a lower level:

1. a lowering of both upper premolar segments;
2. a lowering of the inferior incisive canine segment;
3. an exaggerated opening of the two mandibular angles.

The elevation of the two superior premolar–molar segments is usually carried out by the method proposed by Schuchardt (Fig. 9.5). An inverted 'U' osteotomy, encompassing this dento-alveolar segment, associated with a small supra-apical resection, permits the elevation of this segment, thereby horizontalizing the superior occlusal plane and reducing the open bite at the same

time (Fig. 9.6). The chin is elevated by this procedure. As was shown by Kufner (1960) and later Dautrey (1970), this type of osteotomy can be performed through a single lateral vestibular approach, completely respecting the palatine mucosa, and is therefore carried out in a single operative session. Preliminary model surgery is indispensable in order to assure that the subsequent mandibular autorotation will provide a good incisive occlusal relationship.

In practice, it is quite common for this type of dental malocclusion creating a dystopia of the chin to coexist with a dysmorphia of the chin which in most cases is a hypsogenia with a retrogenia (Fig. 9.7a and c). The diminution in the vertical facial dimension, and the remodelling of the chin are, therefore, obtained by combining a double Schuchardt osteotomy to correct the dental malocclusion with a horizontal basilar osteotomy to reduce the height of the chin and at the same time

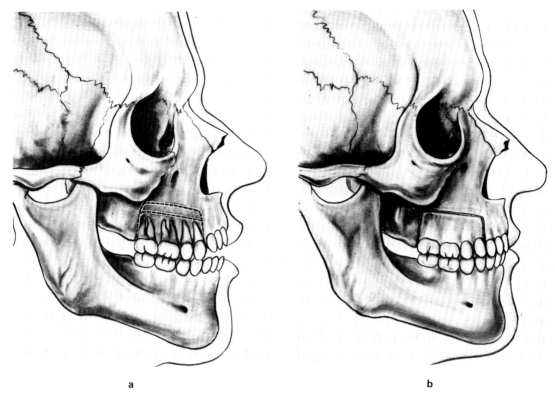

Fig. 9.5   a. Inferior genial dystopia due to an interincisive open bite caused by a lowering of the premolar–molar superior segment. Outline of the Schuchardt osteotomy; **b.** Posterior dento-alveolar elevation reducing the interdental opening, diminishing the facial height and permitting the elevation of the chin by mandibular autorotation.

augment the anterior projection (Fig. 9.7c, d and f).

In more extensive cases, all the preceding anomalies can be found together, leading to a very substantial excess in facial height, as well as a major deformity in the profile of the chin (Fig. 9.8a, c and e). The amplitude of the anterior open bite is caused both by a lowering of the upper premolar–molar regions, and by a lowering of the inferior incisive–canine segment. This double vertical alteration is still further increased by a significant hypsogenia which further aggravates the accompanying retrogenia. A rigorous anatomical analysis naturally leads to a logical therapeutic plan. Three segmental osteotomies assure the correction of the alveolar anomalies, while two basilar osteotomies reduce the height and the retrusion of the chin (Fig. 9.8b, d and f).

An exaggerated opening of the mandibular angle

(which we propose should be called 'amblygonia' (*amblos* = obtuse, *gonios* = angle)) can also be the cause of an anterior interincisive vertical open bite. The dystopia of the chin that it causes can be reduced by an angular rotational osteotomy, with preservation of the inferior dental neurovascular pedicle. The closure of the angle which permits the superior rotation of the horizontal rami of course necessitates the introduction of an interpositional bone graft (Fig. 9.9). One thus arrives at a very stable reconstruction which can effectively counteract the traction of the suprahyoid muscles, the tension on which has been even further increased by the elevation of the chin. Carrying out this procedure by the buccal approach has its risks, and it is preferable to do it from an external approach.

An analogous movement of angle closure by rotation can also be achieved by a sagittal

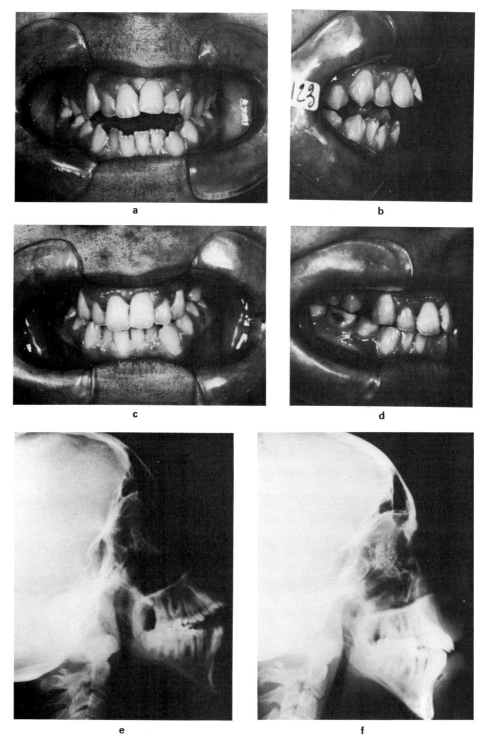

Fig. 9.6   a. and **b.** Anterior vertical open bite as the consequence of a premolar bilateral superior premolar–molar supra-alveolism (Fig. 9.7); **c.** and **d.** Reduction by double osteotomy of the Schuchardt type. Subsequent orthodontic treatment will harmonize the dental alignment. **e.** and **f.** Pre- and postoperative cephalometric films.

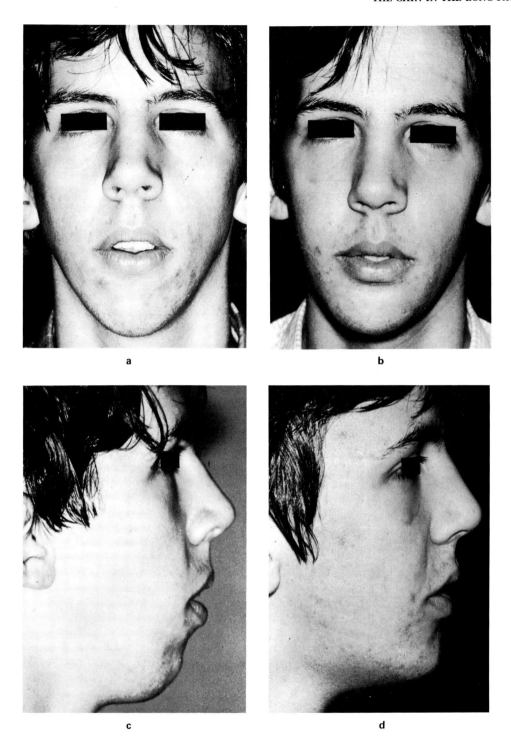

Fig. 9.7 **a.** and **c.** Hypsoprosopia seen in frontal and profile view (interincisive anterior open bite, hypso- and retrogenia); **b.** and **d.** Reduction of the facial height and remodelling of the chin.

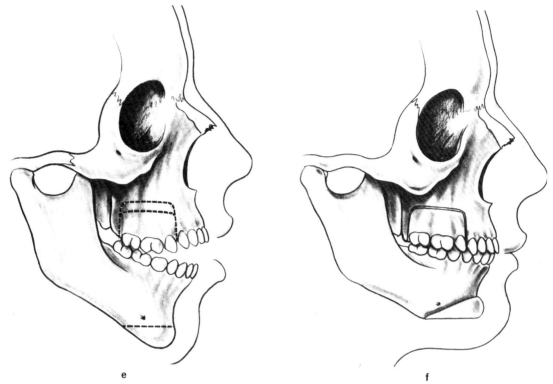

e   f

Fig. 9.7   e. and **f.** Schematic representation of the skeletal anomaly, the proposed osteotomies and the effect of the predicted correction.

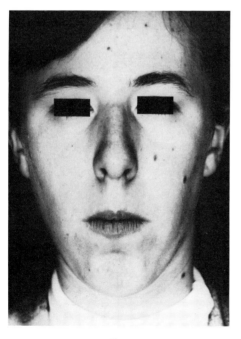

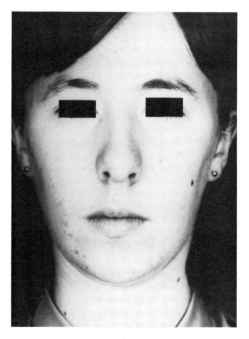

a   b

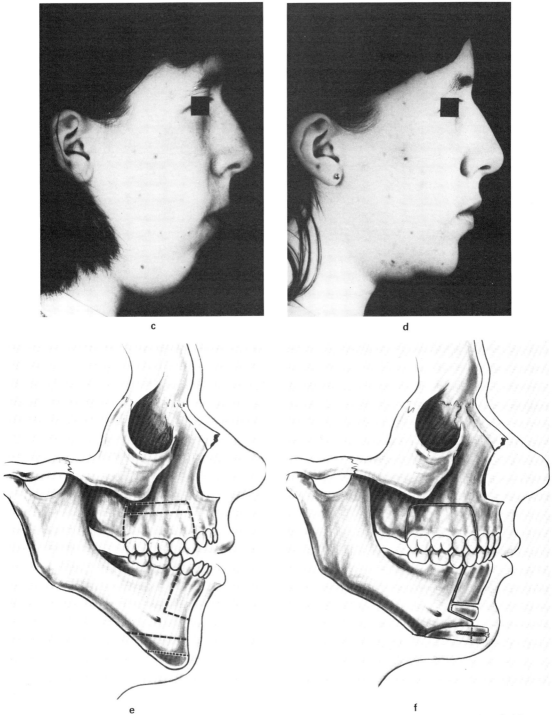

Fig. 9.8 **a.** and **c.** Significant *hypsoprosopia* with retrogenia, in frontal and profile view; **b.** and **d.** Diminution of the facial height and remodelling of the mental contours in the frontal and profile views (the submental fold could have been avoided by a better muscular reinsertion); **e.** Schematic representation of the underlying skeletal lesions: superior premolar–molar lowering, inferior incisive–canine lowering, hypsogenia, retrogenia, and the outlines of the five corrective segmental osteotomies; **f.** Schematic view of the segmental displacements and their effect on the profile of the covering tissues.

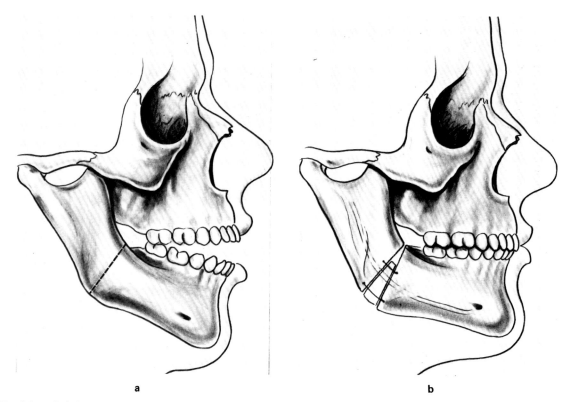

a   **b**

Fig. 9.9   a. Inferior genial dystopia caused by a vertical open bite resulting from an amblygonia; **b.** Angular osteotomy with conservation of the neurovascular pedicle permitting reestablishment of the dental occlusion and reduction in the facial height. Maintenance of mandibular rotation and elevation by placement of an intercalary bone graft fixed by a double osteosynthesis.

osteotomy of the ascending rami. It has, of course, the enormous advantage of being carried out without an external skin incision. However, it involves a substantial posterior–superior protrusion of the internal fragment of the ascending ramus, and the muscular tension on the superhyoid which this causes is difficult to control with certainty by an osteosynthesis, short of a double transfixing screw technique.

This type of amblygonia, creating an intradental open bite and therefore a hypsoprosopia, is in fact rarely an isolated condition; it is frequent to encounter it associated with a retromaxillar anomaly (Fig. 9.10). Its correction is carried out at the same operative session using an advancement LeFort 1 osteotomy. This is done at the beginning of the operation. The extent of the advancement should be carefully calculated beforehand, since because of the open bite, there is no one ideal dental intercuspidation. Study of the cephalometric film

and the preoperative facial appearance are of great importance, as well as the mounting of dental models on a simulating articulator. In effect, the reduction of the open bite determines the vertical position of the chin, and the advancement of the maxilla determines the degree of its projection.

In those cases where both jaws are completely mobilized during the same operative procedure, it is very useful for the surgeon to have a small intermediate dental splint at his disposal, made out of an acrylic resin on the articulated dental models, and then sterilized. After the LeFort 1 osteotomy, it enables the maxilla to be positioned precisely in its desired position before carrying out the mandibular osteotomy.

This type of anterior vertical interdental open bite causing a genial dystopia can only arise in the premolar region (Fig. 9.11a). The sagittal osteotomy of the horizontal rami proposed by Delaire is the treatment of choice here (Fig.

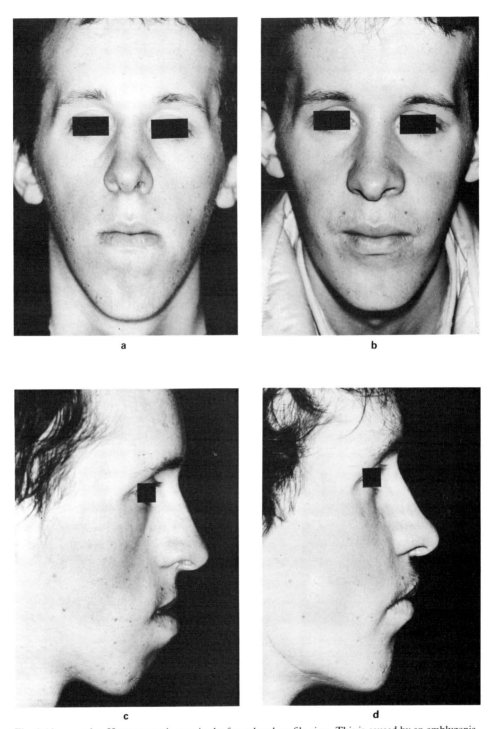

Fig. 9.10   a. and **c.** Hypsoprosopia seen in the frontal and profile view. This is caused by an amblygonia but is accompanied by a retromaxillism; **b.** and **d.** Postoperative views showing reduction of the facial height and remodelling of the chin. Also, one can appreciate that this result would have been better had there been a basilar segmental osteotomy of the chin with elevation and advancement (type shown in Fig. 9.2), but this young patient was quite satisfied and refused further surgery;

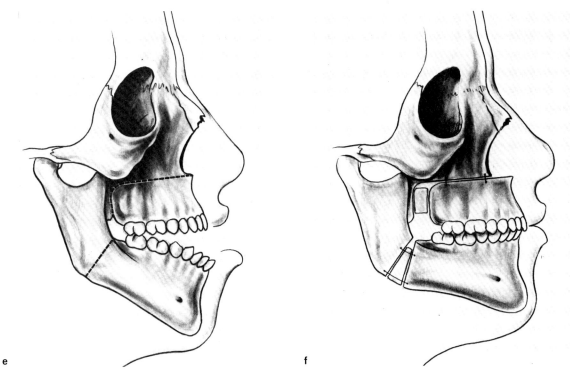

e   f

Fig. 9.10   e. Schematic representation of the underlying skeletal lesions, drawings of the LeFort 1 advancement osteotomy, and the angular osteotomy; **f.** Schematic representation of the underlying postoperative skeletal situation and its effect on the cutaneous contours.

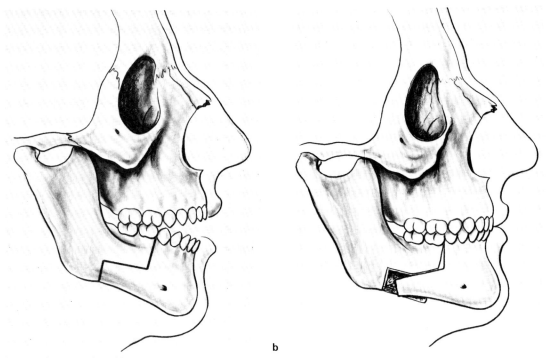

a   b

Fig. 9.11   a. Augmentation of facial height caused by an anterior vertical open bite with its origin in the premolar region; **b.** Reduction of the genial dystopia by reestablishment of the dental occlusion after a sagittal osteotomy of the horizontal rami of the Delaire type.

9.11b). Carried out through an intra-oral approach, without the need for a bone graft, it has the advantage of only requiring stabilization of one osteotomized jaw, which is maintained for a prolonged period and provides a guarantee of consolidation without relapse.

Having dealt with the three main causes of an anterior vertical open bite (amblygonia, an inferior displacement of the lower anterior alveolar segment and an excessive eruption of the upper molar regions) and therefore a dystopia of the chin, there remains one other anomaly to be discussed which can also cause an augmentation of the vertical dimension of the face — the excess maxillary height or the hypsomaxillar anomaly (Fig. 9.12a).

This type of anomaly requires a LeFort 1 type osteotomy for its correction with resection of an intermediate bony segment (Fig. 9.12b). One thus simultaneously corrects the excessive and deforming exposure of the upper incisive–canine segment

below the lower border of the upper lip, as well as the downward displacement of the chin.

As with the previous anomalies, this hypsomaxillar anomaly is rarely found in a pure state, but is much more often associated with other deformities that require correction at the same time. They are located either in the maxilla or in the mandible.

The coexistence of a retromaxillar anomaly does not pose a particular problem in treatment since the movement of elevation effected by the LeFort 1 osteotomy is easily associated with a movement of advancement (Fig. 9.13a, c and e). The association of these two maxillary displacements results in an autorotation in the mandible and, therefore, the spontaneous correction of a vertical anterior open bite when this is necessary (Fig. 9.13f). A noticeable modification in the appearance of the chin results, even though no osteotomy has been done directly on it (Fig. 9.13b and d).

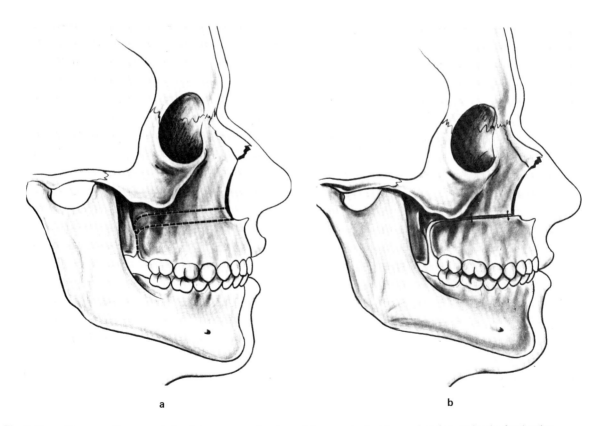

a

b

Fig. 9.12   a. Hypsomaxillar anomaly involving an excessive show of the superior incisives and canines under the free border of the lip. Drawing of the LeFort 1 type elevation osteotomy; **b.** Normalization of the labiomental relationship due to the elevation of the dental arch after resection of the excess maxillary height. The chin has also been elevated.

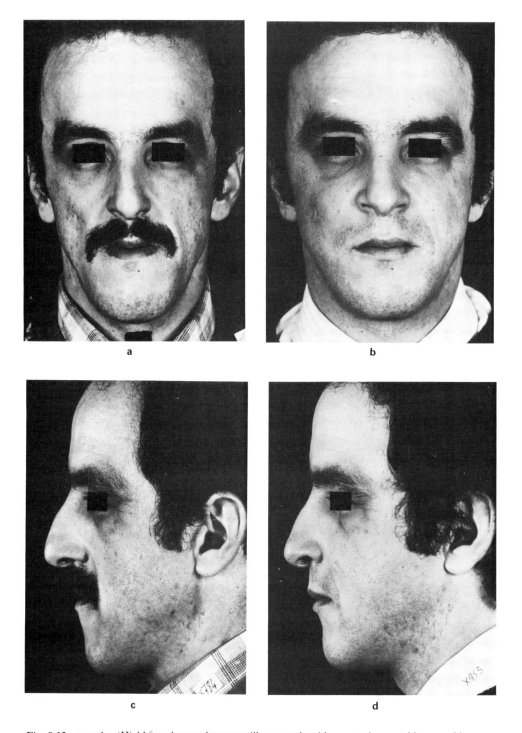

a

b

c

d

Fig. 9.13   a. and **c.** 'High' face due to a hypsomaxillar anomaly with an anterior open bite caused by an amblygonia. Coexistence of a retromaxillar anomaly; **b.** and **d.** Postoperative frontal and profile views;

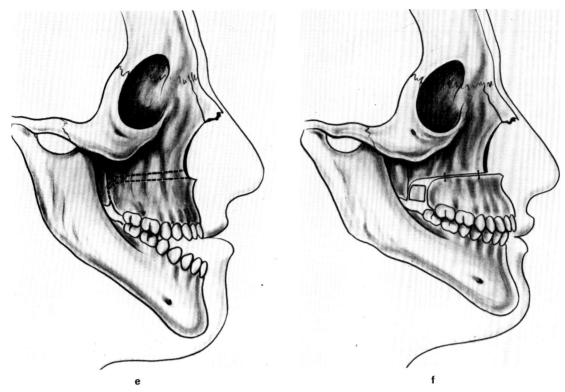

e   f

Fig. 9.13   e. Schematic view of the causative skeletal anomalies. Drawings of the LeFort 1 osteotomy and the resection to correct the hypsomaxillar anomaly; **f.** Elevation of the maxilla combined with its advancement permit a spontaneous autorotation of the mandible which corrects the associated interdental open bite without it being necessary to operate directly in the area of the amblygonia.

On the other hand, the association of a retromandibular anomaly naturally calls for a mandibular osteotomy (Fig. 9.14a and c). According to the requirements of the particular case, this procedure can be carried out on the ascending rami, the angle or the horizontal rami (Fig. 9.14f).

It is certainly useful to carry out the maxillary and mandibular osteotomies during the same procedure. However, to do so requires great precision in developing a treatment plan and evaluating the various movements of the bony segments. A small intermediate splint, prepared on the simulating dental models and then sterilized is in this situation very useful to the surgeon during the operation.

It is not rare that the extent of the required movement of elevation risks causing an interference with the nasal airway. In this situation one can substitute the procedure developed by Bell and Epker for the classic LeFort 1 osteotomy. The maxillary resection is carried out at the level of the palatal vault and not on the lateral wall of the nasal fossae, the nasal floor thereby being respected. Using this method there is no diminution of the nasal airway (Fig. 9.14b, d and f).

In extreme cases, amblygonia, hypsomaxillar and retromaxillar anomalies can exist at the same time (Fig. 9.15a and c, 9.16a and c). These three deformities require simultaneous correction by simultaneous maxillary and mandibular osteotomies. The particular design of the osteotomies depends, of course, on the form of the bony anomaly.

Here too, although no procedure is carried out directly on the chin itself, its form and shape are profoundly modified and one can see a true 'redraping' of the soft tissues around it (Fig. 9.15d and 9.16d).

It should be stated that a rhinoplasty should not be attempted during the same operative

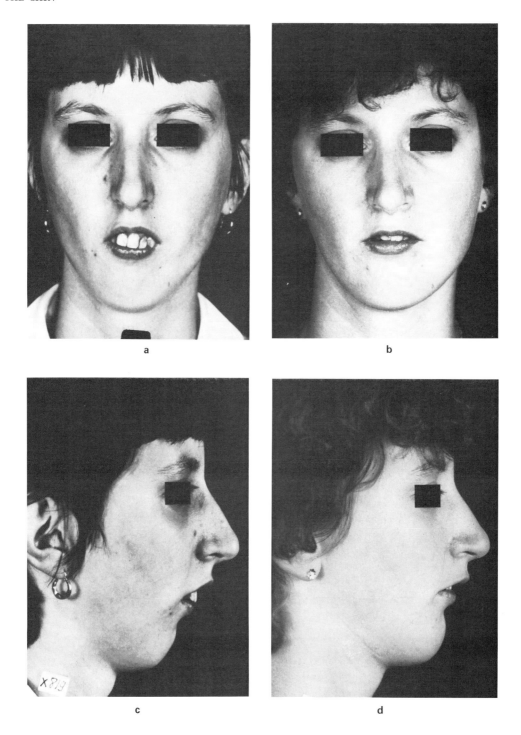

a

b

c

d

Fig. 9.14 **a.** and **c.** Significant hypsomaxillar anomaly involving quite a deforming show of the maxillary incisive–canine teeth and a notable augmentation of the facial height. Associated retromandibular anomaly; **b.** and **d.** Frontal and profile postoperative views. The anteroposterior mandibular correction is still inadequate, and on the frontal view there still persists a slight angular asymmetry;

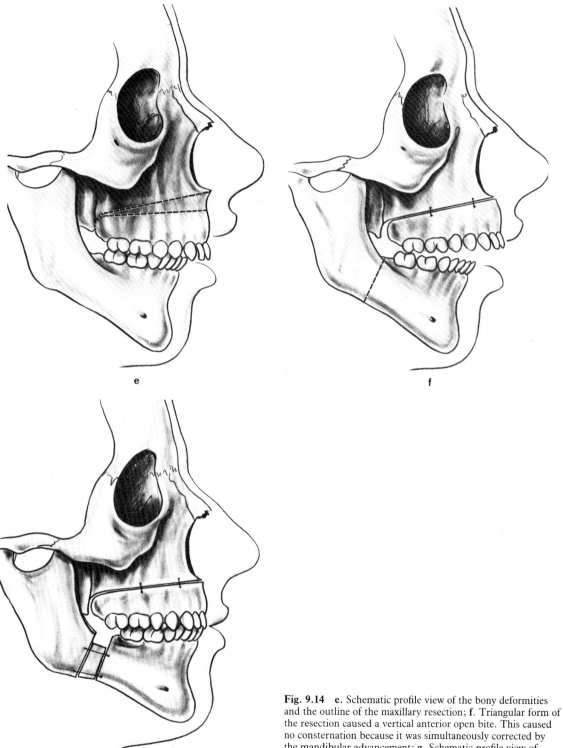

Fig. 9.14 **e.** Schematic profile view of the bony deformities and the outline of the maxillary resection; **f.** Triangular form of the resection caused a vertical anterior open bite. This caused no consternation because it was simultaneously corrected by the mandibular advancement; **g.** Schematic profile view of the final bony relationships and their effects on the facial contour.

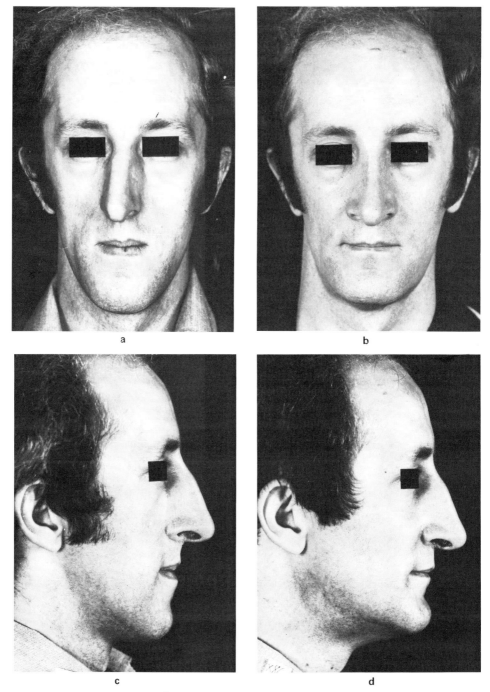

Fig. 9.15   a. and **c.** Genial dystopia with augmentation of the facial height. This is a result of a hypsomaxillar anomaly and an anterior vertical interincisive open bite caused by an amblygonia. It is associated with a retromaxillar deformity; **b.** and **d.** Postoperative frontal and profile views of the face showing the modifications to the form and position of the mental region. There has been a true 'redraping' of the soft tissues over the bony contours, although there has been no osteotomy at this level. As b is taken after the rhinoplasty, d is taken before the rhinoplasty to show the modifications to the profile due solely to the maxillomandibular osteotomies;

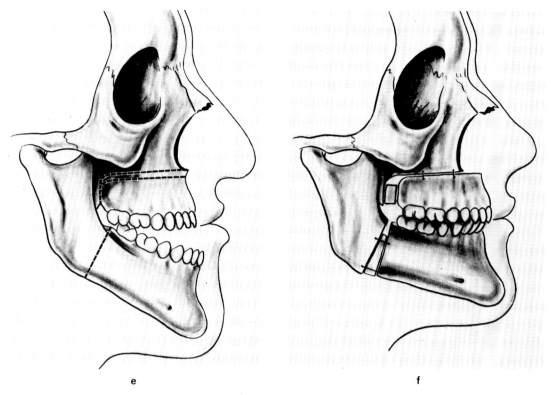

e   f

Fig. 9.15 **e.** Schematic reproduction of the lateral cephalometric film of the skeleton and the soft tissues with the outline of the corrective maxillary and mandibular osteotomies; **f.** Schematic representation of the postoperative cephalometric profile film.

procedure. The evaluation of proportions required by this procedure can be done much more precisely and harmoniously at a second operation, the outer contour of the chin already having been established.

Also, an osteotomy which raises and advances the maxilla has an effect on the profile, on the base of the nose and on the lower portion of the lower nasal dorsum, altering the operative technique that will be required for the eventual nasal correction (Fig. 9.15d).

It also seems preferable to put off an additional complementary segmental osteotomy (similar to that shown in Figs. 9.2e and f) when the combined maxillary and mandibular osteotomies done at a distance have not given an absolutely perfect result (Figs 9.10d and 9.14d), and it is established that there will be the need to reduce a residual hypsogenia and retrogenia. It is much easier to evaluate with precision the amplitude of the required secondary displacements at a later date;

making this judgement in a patient who is already swollen by maxillomandibular surgery, and who is lying flat on the operating table, intubated through the nasal route so that the dental occlusion can be evaluated, is highly risky.

The surgical correction of a 'high' face also has an effect on the profile of the chin.

Surgical correction requires a precise diagnosis beforehand because an increase in facial height is not always caused by a deformity in one area. It can be the result of deformities that are quite varied, and in 1983, along with P. Diner, we proposed a schematic classification.

This latter classification seems quite close to that proposed by H.P. Freihofer for the various anomalies which can cause a diminution of facial height:

1. a hypsogenia or an excess in height of the chin corresponds, in effect, to group 1 of his classification, for the inverse type of anomaly;

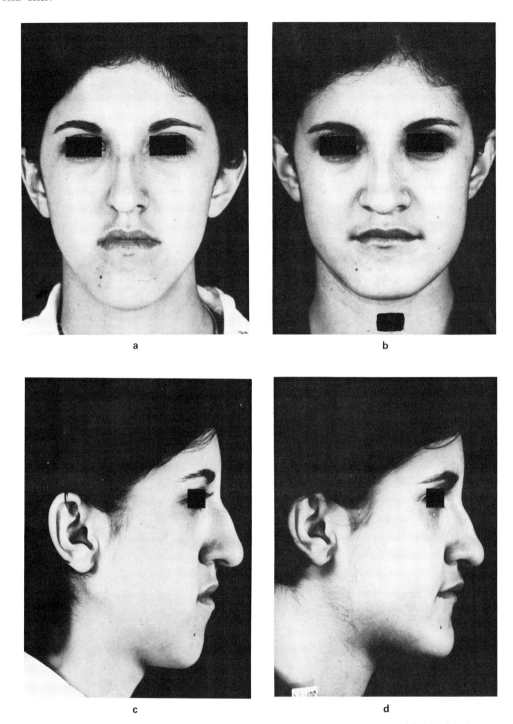

a

b

c

d

Fig. 9.16 a. and **c.** Hypsoprosopia with retromaxillar deformity. The augmentation of facial height is caused by a hypsomaxillar anomaly and an amblygonia (identical lesions to Fig. 9.15e); **b.** and **d.** Postoperative views of the face after maxillomandibular osteotomies carried out in one operation. A rhinoplasty has not yet been done. The outline of the osteotomies is the same as those in Fig. 9.15. Nothing has been done directly in the chin area which has changed notably in its form and position. Note the change in the cutaneous area beneath the angle of the mandible.

2. an anterior open bite, whether it is caused by a lowering of the upper premolar–molar segment, a lowering of the inferior incisive–canine segment or an amblygonia, corresponds to his group 2 represented by an 'over bite';

3. a hypsomaxillar anomaly representing an excess in height in the maxillary infrastructure corresponds to his group 3 describing the opposite type of anomaly.

The architectural analysis proposed by Delaire (1978) has considerably aided the anatomical diagnosis of these vertical facial deformities associated most often with sagittal anomalies. This type of analysis helps to support the clinical diagnosis where, we feel, good sense and eclecticism should always be the artistic inspiration of the surgeon.

The most important task is a rigorous, definitive diagnosis of the deformity. The most appropriate treatment plan then follows logically.

We would like to thank the *l'Encyclopédie Médico-Chirurgicale* and particularly Dr. J. L. Vincent, editor of the *Traité de Stomatologie*, for permitting us to reproduce the drawings which their artist Monsieur D. Duval did for us for their publication.

BIBLIOGRAPHY

Bell W H, McBride K L 1977 Correction of the long face syndrome by LeFort I osteotomy. Oral Surgery, Oral Medicine, Oral Pathology 44: 493–540

Converse J M, Wood-Smith D 1964 Horizontal osteotomy of the mandible. Plastic and Reconstructive Surgery 34: 464

Dautrey J 1970 Personal communication, 17 January

Delaire J 1978 Analyse architecturale et structures cranio-faciales. Revue de Stomatologie et de Chirurgie Maxillofaciale 79: 1–33

Delaire J 1970 De l'intérêt des ostéotomies sagittales dans la correction des infragnathies mandibularies. Annales de Chirurgie Plastique 15: 104

Epker B N, Fisch L C 1978 The surgical orthodontic correction of class III skeletal open bite. American Journal of Othodontics 73: 601–618

Farkas L G, Hreczko T A, Kolar J C, Munro I R 1985 Vertical and horizontal proportions of the face in young adult north american caucasians = revision of neoclassical canons. Plastic and Reconstructive Surgery 75(3): 328–337

Freihofer H P M 1981 Surgical treatment of the short face syndrome. Journal of Oral Surgery 39: 907–911

Garnier M, Delamare V 1985 Dictionnaire des termes techniques de médecine. Maloine, Paris, 21st edition

Koële H 1959 Surgical operation on the alveolar ridge to correct occlusal abnormalities. Oral Surgery, Oral Medicine, Oral Pathology 12: 413–515

Kufner J 1960 Nove metody chirurgickato leceni otevreneho skusu. Stomatologiia 60: 387

Merville L C 1980 Béance verticale antérieure intermaxillo-mandibulaire. Réduction chirurgicale secondaire. Ed. Prélat. Orthodontie Francaise 51: 425–440

Merville L C, Diner P A 1984 Dysmorphies verticales maxillo-mandibulaires par excès. Revue de Stomatologie et de Chirurgie Maxillofaciale 85(1): 12–22

Obwegeser H 1968 Die Bewegung des unteren alveolar Fortsatzes zur Korrectur von Kieferstellungs nomaliari. Zeitschrift für Zahnärztliche Orthopädie 23: 1075–1084

Obwegeser H 1969 Die Bewegung des unteren alveolar Fortsatzes zur Korrectur von Kieferstellungs nomaliari. Zeitschrift für Zahnärztliche Orthopädie 24: 5–15

Obwegeser H 1957 The surgical correction of mandibular prognathism and retrognathie with consideration of genioplasty. Oral Surgery, Oral Medicine, Oral Pathology 10: 677

Schuchardt K 1960 Experiences with the surgical treatment of some deformities of the jaws. In: Transactions of the International Society of Plastic Surgeons. Livingstone, Edinburgh and London, pp 73–78

10. The chin in temporomandibular ankylosis

F. Souyris

Temporomandibular ankylosis and traumatic or infectious lesions of the mandibular condyle lead to an arrest of facial growth. The mechanism most often invoked is an interference with the activity of the condylar ossification centre, but difficulties with muscular dynamics are also involved in these deformities. These are well known and the deformity of the chin is only an element of a more complex dysmorphosis which involves not only the mandibular level but also the mid-portion of the face. For the deformities to appear clearly, it is necessary for the arrest of condylar activity to occur early in facial development, which is to say for the process to affect infants less than four years of age.

CLINICAL DESCRIPTION

Although the deformity involves the entire face, it is in the chin that it manifests itself the most clearly and that it has its most characteristic expression. Different clinical pictures are noted depending on whether the ankylosis is unilateral or bilateral.

1. In the unilateral forms, the examination of the face shows a deviation of the point of the chin (which we take to be the lowest point of the mandibular contour) towards the affected side. The chin is like an 'accusing finger' pointing at the affected condyle (Figs. 10.1, 10.2 and 10.3).

When the median crease is marked, it follows this movement and becomes located adjacent to the midline. This gives the impression that the mental eminence on the affected side is more marked and projects further than its homologue. In the vertical plane, the mental eminence is lower on one side than the other.

If one examines the paramedian region one

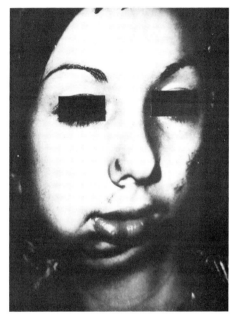

Fig. 10.1 Characteristic and quite marked deformity from a right temporomandibular ankylosis.

notes that on one side it appears normal with a convex curvature and on the other side it is flattened and sometimes gently concave. Additionally, in the more pronounced forms one notices a sort of notch. This is a trap for the non-specialist who thinks that the affected side is the one where this flattening occurs, whereas in reality this is a characteristic of the healthy side.

Examination of the profile shows a retrusion of the chin less marked than in the bilateral forms but present nevertheless. In addition, the height of the upper portion of the face is diminished in a way that the lower lip becomes everted with a very marked sublabial crease.

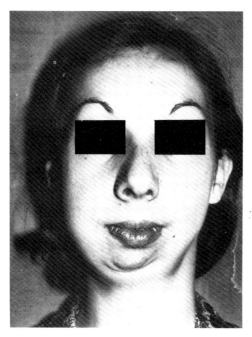

Fig. 10.2 After correcting the ankylosis, the asymmetry of the chin has improved slightly due to a muscular rehabilitation.

2. In the bilateral forms, the typical retrusive type of the 'Vogelgesicht' (bird face) exists and the chin is the main characteristic of the deformity (Figs. 10.4, 10.5, 10.6 and 10.7).

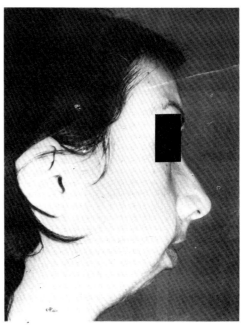

Fig. 10.4 Retrusive profile of an ankylosis which occurred in early infancy.

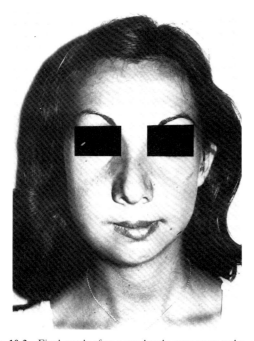

Fig. 10.3 Final result after a quadruple osteotomy and a chin implant.

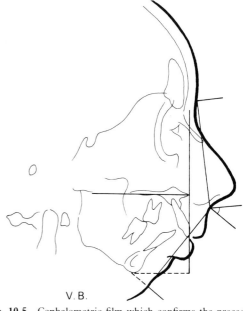

V. B.

Fig. 10.5 Cephalometric film which confirms the preceding photograph.

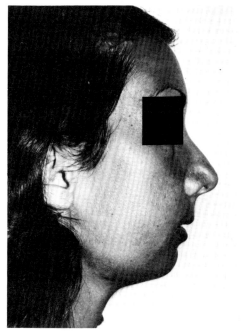

Fig. 10.6 Result after a mandibular advancement with continuous postoperative forward traction and a later chin implant.

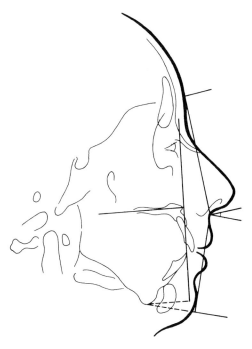

Fig. 10.7 Cephalometric film confirming the preceding photograph.

There is certainly a retrusion, but there is also an atrophy and a diminution of height. There is no longer a mental eminence, and the inferior border of the symphysis is more posteriorly placed than the superior border. The lower lip is compressed by the upper lip which dominates it and everts it. The fleshy lip is everted with a crease that is so marked that it becomes a true wrinkle due to the destruction of the elastic fibres.

In the very accentuated forms, for example when there has been an osteomyelitis in a young infant which has resulted in a mandible reduced to a thin rod, the chin disappears and the cervical profile is continuous with that of the upper lip and the columella.

The clinical workup should be completed by photographs and a cephalometrical analysis:

a. Photographs. Frontal and profile, they complete and document the clinical workup. It is not necessary to trace reference points on the photograph; this can be reserved for the cephalometrical film. On the other hand, the head position and the centering of the subject on the photograph should be perfect.

b. Cephalometrical analysis. The study of the chin becomes part of a three dimensional study which can be obtained by teleradiographic cephalometrical films taken under the usual conditions. For the study of the profile one should include the soft tissues; in this way one can predict the postoperative result as it will appear on the facial profile.

The lateral facial profile is essential to define a surgical technique, which is to say, to choose between an operation carried out on the bony chin only and an operation carried out on the mandible as a whole. In sum, this is the decision between a genioplasty and a mandibular osteotomy.

TREATMENT

The therapeutic approach is a function of the extent of the malformation, the degree of the asymmetry or atrophy of the mandible, and the participation or lack of participation of the middle third of the face.

The age at which treatment is given is the object of much discussion and consensus has not been reached at the present time. The choice between an early operation and an operation after the cessation of growth is less of a question for the chin than for the facial dysmorphosis taken as a whole, and interceptive operations to lengthen the ascending ramus have most of their effect on the mandibular asymmetry rather than on specific correction of the chin which remains deformed. Thus one is often obliged to return later to this area.

We prefer to treat the deformity of the chin either by itself or as part of a more complex number of facial osteotomies, but always after the cessation of growth.

1. In the unilateral forms, if the asymmetry is confined to the chin, without the participation of the middle portion or the body of the mandible, we will perform a mentoplasty by slightly oblique section of the symphysis and translation of the basilar fragment of the chin from the ankylosed side towards the unaffected side (Obwegeser & Trauner 1957). This type of operation is helped by the technique described by Neuner & Meyer (1976), which consists of carrying out the correction in two or three separate stages, as shown in fig. 10.8.

If there is facial asymmetry, we carry out a quadruple osteotomy (Souyris 1975). The chin is treated at the same time as the rest of the face.

In certain cases, after such a procedure or an adjacent procedure, there remains a certain degree of retrogenia and in this situation an implant is justified — either a silicone prosthesis or an implant of a coral derivative.

2. In the bilateral forms, if there is a slight retrogenia, one carries out a horizontal mentoplasty with an advancement of the symphyseal fragment, or a prosthetic implant.

If the retromandibular anomaly is associated with a retrogenia, it is necessary to carry out a mandibular osteotomy:

— either of the horizontal ramus with a basilar graft or an advancement in drawer fashion of the anterior fragment (Ginestet et al 1955);
— or of the ascending ramus through a cutaneous approach with an 'L'-shaped osteotomy and insertion of a bone graft.

If the deformity is more marked with an extreme retromandibular anomaly, we perform a two stage lengthening:

— at the first stage, an osteotomy of the ascending ramus through a cutaneous approach, and application of a traction device which is attached to a head frame through a traction wire on the symphysis;
— at the second stage (three weeks later) after

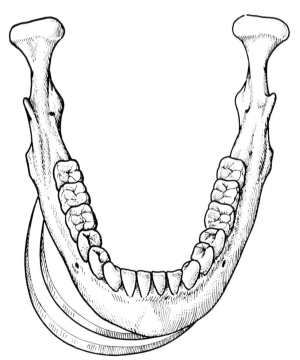

Fig. 10.8   The technique of Neuner described in the text.

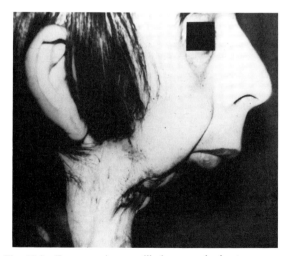

Fig. 10.9   Extreme micromandibular anomaly due to mandibular osteomyelitis in the newborn.

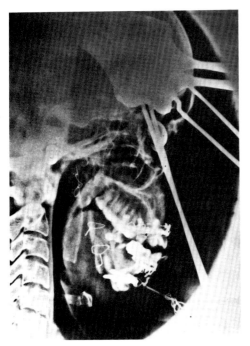

Fig. 10.10 Treatment by osteotomy of the ascending rami with continuous traction on the symphysis and an interpositional bone graft in the area of lengthening.

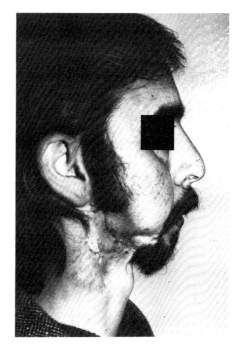

Fig. 10.11 Result after a chin implant.

elongation and even hypercorrection, the fragments are fixed with an autogenous bone graft. As in the other cases, if a residual retrogenia persists, this is corrected by an implant (Souyris et al 1975) (Figs 10.9, 10.10 and 10.11).

CONCLUSION

The deformity of the chin in temporomandibular ankylosis is well known and its correction requires the proven and classical techniques which were presented long ago by Ginestet, Trauner and many others.

More recent progress came in the 1970s with techniques of osteotomy involving the entire facial skeleton.

The chin is only one element of the deformity and is not the exclusive location of it. A treatment which thus deals with the underlying causes more than with the external appearances gives results that are more stable and more harmonious.

BIBLIOGRAPHY

Ginestet G, Dupuis A, Merville L, Gorek E 1955 Traitement chirurgical d'une micrognathie très accentuée associée à une pro-alvéolie supérieure. Revue de Stomatologie 56: 834
Neuner O, Meyer W 1976 Ostéotomie horizontale dans le traitement des latéro-mandibulies. Revue de Stomatologie 76(1): 105–107
Obwegeser H, Trauner R 1957 The surgical correction of mandibular prognathism and retrognathia with consideration of genioplasty. Surgical procedures to correct mandibular prognathism and reshaping of the chin. Oral Surgery, Oral Medicine, Oral Pathology 10: 677
Souyris F 1975–1976 Surgical treatment of facial asymmetry. Transactions of the VIth international congress of Plastic and Reconstructive Surgery. Masson, Paris.
Souyris F, Brami S, Caravel J B , Ganigal A 1975 Traitement d'un cas d'extrême micrognathie. Intérêt l'extension continu. Revue de Stomatologie 76(1): 39–44

11. Chin augmentation

G. Aiach

The choice of a method of chin augmentation can, at first, seem simple. In fact, it depends on the training of the surgeon and the environment in which this surgery is carried out.

In the absence of maxillofacial training the choice is simple — one uses prostheses; small if the retrogenia is discrete and more sizable if it is significant. The submental approach is the one most often used, perhaps for fear of infection. Results can certainly be excellent but the indications for an osteotomy are sometimes present.

On the other hand, certain maxillofacial surgeons never use a prosthesis even in the most minor cases. This is explained in part by the fact that they only rarely see those cases of discrete retrogenia which present themselves in the consulting room of an aesthetic surgeon, where the correction is proposed to a patient who consults the surgeon for something other than their chin (rhinoplasty, face lift).

SURGICAL APPROACHES

I. The transmucosal approach

This avoids any visible scar and provides a rapid route to the mental symphysis. But it demands several precautions:

a. Before surgery there needs to be very good intra-oral hygiene (thorough cleaning of the teeth 15 days before the procedure), and one should verify that there are no cysts at the apices of the incisive and canine teeth.
b. The surgical preparation should be fastidious (cleaning with surgical soap or brushing), and on the day of surgery there should be a cleaning of the dental spaces.

c. After surgery, there needs to be local care, brushing, and mouth washes.

All these precautions should provide a very low incidence of infection whichever procedure is used.

After the intra-oral approach, there can be considerable pain calling for analgesics and also a temporary diminution in the mobility of the lower lip which modifies the smile and may bother the patient.

2. The transcutaneous approach

Its only inconvenience is a scar, which is in fact not very visible since it is located in a zone of shadow.

This provides an approach to the inferior border of the mental symphysis which takes longer when one compares it to the extreme rapidity of the vestibular approach.

3. Indications

Indications for one or the other approach depend on the desires of the patient, on the habits of the surgeon and on the extent and type of the deformity. In simple terms, one can choose

The cutaneous approach:

a. for the discrete retrogenias (an incision of 10–15 mm is sufficient);
b. if the body of the mandible is quite short with a deep gingivolabial crease; (In this case, the dissection of the pocket can be more precisely performed in its upper extent, making it possible to avoid the protrusion of the prosthesis into the buccal vestibule. According to certain authors, large

implants are more easily placed through a cutaneous approach.)

c. If a submental lipectomy is to be performed at the same operative procedure.

The transmucosal approach:

a. in cases where an osteotomy is to be carried out for which it is indispensable;

b. for the placement of a silicone prosthesis (used quite frequently).

PLACEMENT OF A SILICONE PROSTHESIS

The transmucosal approach (Fig. 11.1)

The choice of the implant

The different models proposed by the manufacturers vary in size, thickness (7–9 mm at the midpoint) and their extremities which may be more or less tapered.

Silicone, having a firm consistency, has the advantage that it can be modified or cut, notably in cases of excessive height (more than 10 mm), where it is necessary to reduce a horizontal section by one third with a scalpel (Fig. 11.2).

The mid portion of the prosthesis sometimes has a notch or indentation of varying depth whose purpose is more to identify the midline than to create a chin dimple.

If this notch is not clear, it is useful to make a mark with a scalpel by cutting away a small piece of the implant; this makes it possible once the implant is in place to move it slightly to the right or to the left and get it into the exact midline.

The implant should be manipulated with the greatest precaution and placed inside a metal cup rather than a compress (one should also avoid manipulating the implant with surgical gloves covered with talc).

Placement of the implant (Fig. 11.3)

1. The cutaneous landmarks are traced with ink:

   a. on the midline;
   b. along the outline of the prosthesis whose lower border should follow but not extend beyond the lower border of the mandible and the mental symphysis.

2. The mucosal incision is outlined. Whether this is traced at the midline with an inverted 'V'

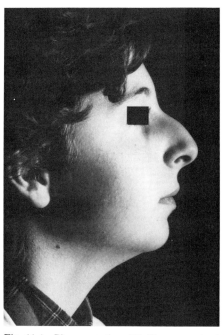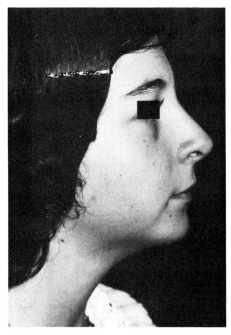

Fig. 11.1 Placement of a solid silicone prosthesis by an intra-oral approach associated with liposuction of the submental area.

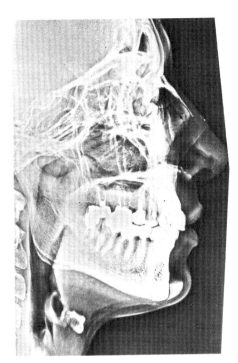

Fig. 11.2 Cephalometrical film showing a silicone prosthesis in place. The height of the prosthesis has been reduced by one third. The position is good, in front of the pogonion.

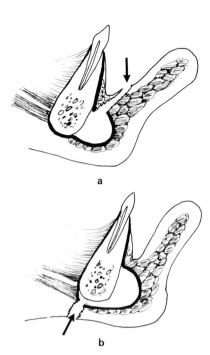

rig. 11.4 a. The mucosal approach: at the upper limit of the pocket, one notes that some muscular tissue has been preserved from the alveolar bone. **b.** The submental cutaneous approach: the periosteal incision is situated behind the lower symphyseal border.

shape or laterally (we will see later the advantages of this latter approach), it is always carried out at a distance from the gingivolabial crease and, therefore, the lip. Its length varies between 15 and 20 mm. After infiltration with a xylocaine adrenaline solution, and incision of the mucosa, the scalpel approaches the muscular plane at a 45 degree angle relative to the plane of the symphysis in order to avoid incising the periosteum more inferiorly and to leave a thickness of muscular tissue

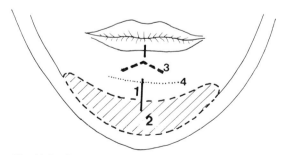

Fig. 11.3 Cutaneous tracings of the midline (1) and the extent of undermining (2). Relationships between the zone of undermining and (3) the mucosal incision and (4) the labiogingivo fold.

adherent to the alveolar bone, which makes it possible to carry out a muscular layer suture and prevent displacement of the implant superiorly (Fig. 11.4).

After the horizontal mucosal incision, Millard proposes vertically incising the muscle which is closed once the prosthesis has been put in place.

3. Preparation of the pocket. The principle is to create a median symmetrical pocket whose limits correspond to the cutaneous outline. This pocket should not be too large in order to avoid secondary displacement, nor should it be too small.

The periosteum is dissected laterally, only extending slightly beyond the limits of the cutaneous markings and down to the inferior border of the bone without rupturing the periosteum, particularly laterally. Periosteal elevators, curved to the right and to the left, can be used in order to keep to the superolateral limits of the undermining, but with practice, the same elevator can carry out all of the undermining. The superior extent of the pocket should be located below the dental apices.

If one wants to place the prosthesis above the periosteum, the undermining is carried out supraperiosteally over the midline for a distance of 2 to 3 cm but laterally, through small, vertical incisions made with the periosteal elevator, the undermining passes into a subperiosteal plane into which the ends of the prosthesis are placed. In effect, over the midline, the soft tissues are quite thick and the prosthesis is rounded; there is therefore no risk here of having an angular protrusion. Laterally, however, the soft tissues are quite thin and the prosthesis can sometimes cause a protrusion whether its ends are tapered or blunt, which explains why a bad position of the prosthesis at this level can cause a subcutaneous projection of its lower border. The supraperiosteal position of the prosthesis makes it possible to limit secondary bony resorption. (Fig. 11.5).

4. Placement of the prosthesis is simpler if the incision is large, but an excessively large incision increases the risk of displacement.

The advantage of a short incision is that it creates a pocket the superior and lateral borders of which are sufficiently stretched or tight to prevent superior displacement.

It can be simpler to place the prosthesis when an incision is made laterally; thus, the largest portion of the prosthesis is placed in this lateral incision, the other extremity being easily guided into place by an Aubry or Aufricht retractor.

The prosthesis should be placed in a symmetrical, median position and maintained there.

The appreciation of the median position is difficult if a notch has not been made on the prosthesis. The end of the periosteal elevator placed in the notch makes it possible to centre the

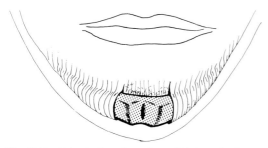

Fig. 11.5 Only the lateral portions of the prosthesis go under the periosteum. The median third is in front of the periosteum.

prosthesis perfectly, moving it gently side to side and aligning the notch with the central inter-incisive space.

If the pocket is too large and the prosthesis seems too mobile within it, one should not hesitate to anchor its mid-portion to the skin by use of a transcutaneous suture tied over a pledget.

The curve of the bony chin is sometimes less open than that of the prosthesis which can cause a retrusion of the end of the prosthesis in spite of the pressure of the soft tissues; in these circumstances it is useful to section the prosthesis vertically and almost completely through at its mid-portion (at the level of the notch).

Suturing is done in two layers: with catgut in the muscle and silk in the mucosa. A moderately compressive dressing is applied directly to the skin using elastic bandages.

Strict buccal hygiene is advised to the patient and it is not necessary to prescribe antibiotics.

The cutaneous approach (Fig. 11.4b)

This is indicated in the case of discrete retrogenia when the resected tissues from a septorhinoplasty are used for the implant or when the mandible is very short.

The horizontal skin incision is made slightly behind the lower symphyseal border. The periosteum is incised behind the inferior symphyseal border and then undermined. If one wants to place the implant subperiosteally, this undermining is more difficult than with the transmucosal approach.

Suture is always carried out in several layers. The periosteal layer is sutured when possible to serve as a barrier against eventual inferior displacement of the implant, and then subcutaneous tissues and skin are sutured.

Fragments of septal cartilage, and the nasal hump, taken in one piece, can also be placed through this approach after having removed all epithelial tissue. Constructions superimposing several fragments and held together by transfixing sutures or stacked up can also be used in order to obtain a graft which follows the form of the bony chin without particularly marked irregularities (Fig. 11.6).

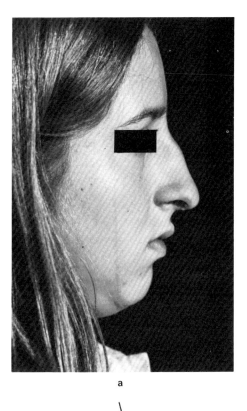

a

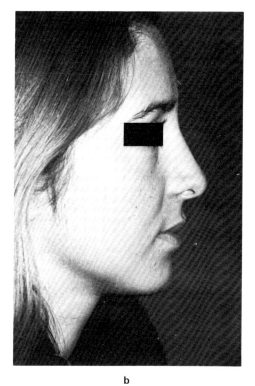

b

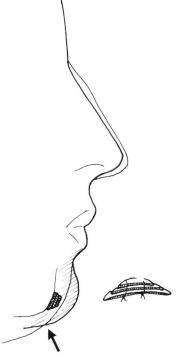

c

Fig. 11.6 Chin augmentation by a submental cutaneous approach: the nasal hump has been removed en bloc and placed on two thicknesses of septal cartilage. This entire fabrication is then held together with three catgut sutures. The postoperative result at one year is shown in **b**.

THE ADVANTAGES AND DISADVANTAGES OF SILICONE PROSTHESES

I. The advantages

These include the great simplicity of the technique, the rapidity of insertion between 5 and 15 minutes and the possibility of carrying out the operation under local anaesthesia by infiltrating in the area of the incision and at the mental foramen.

II. The disadvantages

The main disadvantages are not that the lip is immobile for several weeks when the intra-oral incision is used, or the rare cases of infection (one case in 123), but rather the displacement of the prosthesis seen in the immediate postoperative period.

Displacement is most often associated with an asymmetrical or excessively large pocket.

A vestibular protrusion occurs if the prosthesis is placed too high with respect to the alveolar bone or when the height of the mandible is short; in this latter case, it is preferable to reduce the height of the prosthesis.

A protrusion underneath the skin, often lateral at the level of the inferior border, may be seen where the skin is thin and where a poor fit of the prosthesis to the curve of the bone can cause an absence of continuity of the lower border. This protrusion can also be observed in the mid-portion, immediately adjacent to the inferior border of the mental symphysis, when the cutaneous approach is being used but when the pocket is insufficiently large and a periosteal layer is lacking.

An asymmetry can be observed if the prosthesis has been poorly centred or if it is too high on one side or the other.

Osseous resorption beneath the implant has been the study of many publications which are cited in the further reading list; many authors have studied this phenomenon and tried to find a means of avoiding it. It is rather disagreeable for the surgeon to discover that several years after the operation there is a bony rectangular notch into which the prosthesis seems to have recessed. This causes no harmful effect if the prosthesis is not in contact with the alveolar bone and the dental apices, but at the level of the mental symphysis.

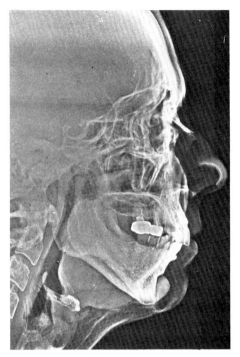

Fig. 11.7 Bony resorption underneath a chin implant which was placed a bit high.

These bony resorptions are only discovered if a systematic study is being done or if a lateral X-ray is carried out for some other cause (Fig. 11.7.).

The resorption is increased if the prosthesis is placed too high in relation to the alveolar bone, or if the pocket is tight; although in certain cases the prosthesis 'floats' in its pocket, and in spite of an extended undermining it seems to be narrowed and pressed against the symphysis by the pressure of the soft tissues. This tension of the soft tissues can be associated with a labiomental dysfunction which should be diagnosed beforehand and provide an indication for an osteotomy.

Bony resorption is often limited to 3–4 mm, which is about half of the thickness of the prosthesis. Its very gradual appearance explains why the recession occurs unnoticed by the patient.

Can one prevent bony resorption?

One can rather limit it:

a. By reserving the indication for cases of discrete retrogenia, and in consequence avoiding using prostheses which are excessively large.

b. By placing the prosthesis in front of the periosteum in its medial one third and under the periosteum laterally.

c. By carrying out several periosteal incisions through the anterior surface of the pocket when this seems to be tight.

THE SUPRABASILAR OSTEOTOMY

This is the simplest osteotomy of the face; it nevertheless requires a certain experience in intra-oral surgery and osteotomies. Described by Trauner and Obwegeser in 1957, it includes a variety of techniques which are adapted to the particular deformity.

The vestibular incision (Fig. 11.8) A preliminary marking is made about 10 mm on the lip side of the gingivolabial fold, extending from canine to canine, and leaving the inferior frenulum behind.

Infiltration by a xylocaine or adrenaline solution is carried out along the incision, then towards the mental symphysis, laterally below the mental pedicles and across the posterior surface of the symphysis. The volume injected is in the region of 10 to 15 ml.

The incision is first mucosal, the lip being held by two hooks, and the scalpel oriented obliquely in relation to the anterior surface of the body of the mandible, sectioning the underlying muscular plane and arriving at the periosteum a little above the level of the line of the osteotomy. This facilitates the suturing of the wound.

The dissection (Fig. 11.9) is carried out with a periosteal elevator in the subperiosteal plane, extending below to the inferior border of the mental symphysis and posteriorly to the angles of the mandible which permits a better redistribution of the soft tissues after the correction; the mental pedicles should be isolated and freed. This extensive undermining makes it possible to obtain a good exposure and a true 'de-gloving' of the symphysis using a special retractor which hooks beneath the symphysis and retracts the soft tissues, whilst two protective retractors are placed laterally.

A precise marking of the osteotomy line is then carried out: in its mid-portion the line is located below the dental apices, 8 to 10 mm above the inferior border of the symphysis; laterally it passes immediately beneath the mental pedicles and reaches the basilar border quite far to the posterior. The direction of this line is determined by the nature of the deformity.

The osteotomy (Fig. 11.10) is preferably carried out by an electric saw which permits a clean, rapid section, a particularly smooth and flat cut, which facilitates the sliding of the inferior fragment with the best possible contact.

A preliminary vertical scoring line is made on the outer cortex to establish the midline, then a

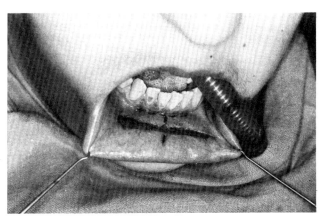

Fig. 11.8 Vestibular incision. This is carried out at some distance from the gingivolabial crease (dotted line). The midline is marked by the scalpel.

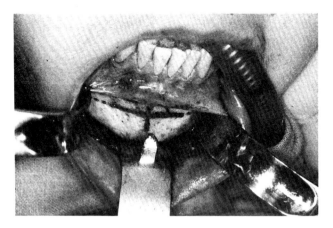

Fig. 11.9   Dissection and exposure. A special elevator is placed
beneath the symphysis; two Dautrey retractors are placed laterally.
The horizontal line of the osteotomy and the midline are marked.

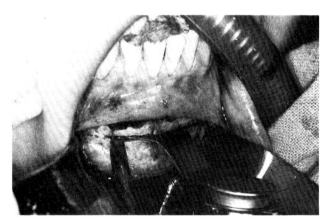

Fig. 11.10   Horizontal osteotomy: medial segment. Having marked
the midline with a small vertical notch, the horizontal osteotomy is
carried out in its median portion with an oscillating saw.

section strictly horizontal to this is carried out
using the oscillating saw in the symphyseal region
where the bone is the thickest.

The osteotomy (Fig. 11.11) is continued laterally
using an alternating saw and the mental pedicles
are protected by a straight retractor which is
applied to the lower border. A perfect symmetry,
between the two lines of the lateral should be
obtained.

If an intermediate geniectomy is necessary, this

is carried out at this stage by performing a second
section superiorly which makes it possible to
remove an intermediate fragment the cuts of which
are parallel or slightly wedge-shaped.

Mobilization (Fig. 11.12) of the inferior
fragment is carried out with the use of fine
osteotomes which make it possible to complete the
fracture at the level of the inner table and then to
displace the mental fragment inferiorly.

One then checks that the edges of the section

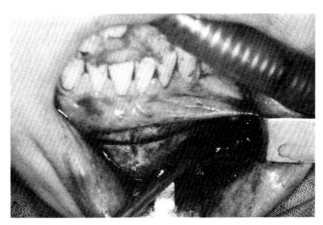

Fig. 11.11 Horizontal osteotomy: lateral segments. The osteotomy is continued laterally with a straight reciprocating saw, the mental pedicles being protected.

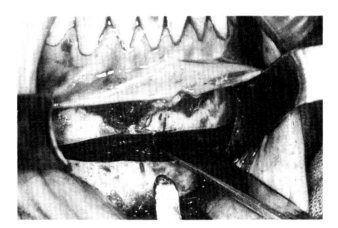

Fig. 11.12 Mobilization of the inferior fragment. After the osteotomy, the mobilization is completed by a thin osteotome which displaces the inferior fragment. One should check that there are no spicules along the osteotomy lines.

do not have spicules or irregularities, particularly posteriorly where the surfaces are in contact after sliding forward.

Once the inferior fragment has been well mobilized (Fig. 11.13) one establishes that the undermining is laterally complete on both sides by successively exteriorizing each half of the mental fragment which pivots on its posterior median pedicle. The periosteal elevator makes it possible to complete the undermining on each side except in the area of the genial apophysis where a geniohyoid muscular pedicle is partially conserved.

In extreme cases, it is possible to reduce this pedicle by stretching it or even by dividing it. However, preservation of the pedicle, even if it is only partial, provides a better bony vascularization and a better contour of the submental region.

The anterior sliding (Fig. 11.14) of the mental fragment can then be carried out; the advancement of the fragment is maintained by a bone forcep and one checks for good coaptation of the surfaces in

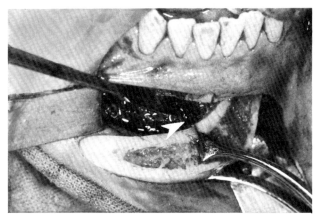

Fig. 11.13 Conservation of a narrow median muscular pedicle. The arrow shows the preserved median pedicle, as well as the lateral portions of the mobilized osseous fragments, all soft tissues having been removed laterally.

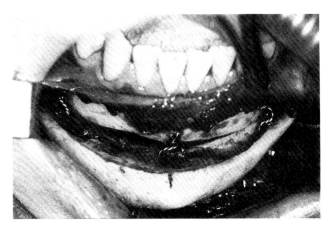

Fig. 11.14 Anterior slide and osteosynthesis. After the advancement of the inferior fragment, an osteosynthesis wire is placed at the midline and two further wires are positioned laterally.

contact before fixation with stainless steel wire; to have a good coaptation it is sometimes necessary to burr down the bony surfaces.

The osteosynthesis is provided by three 0.4 mm wires, one placed medially and the other two laterally. The perforations are made through the outer cortex of the mandible and through the inner cortex of the mental fragment; the first median wire is passed, followed by the two lateral wires, at the level of the mental foramen. Once the segment is well positioned, the first median wire is

tightened and then the lateral wires. One thereby obtains a very stable and solid fixation.

Remodelling (Fig. 11.15) of the advancement after stabilization can be carried out by using a large acrylic burr or a bone burr; any sharp edges, irregularities or protrusions which can be seen laterally along the basilar border can be levelled off. Palpation through the skin of the osseous contours makes it possible to carry out this bony remodelling better.

The soft tissues of the chin, because of the ex-

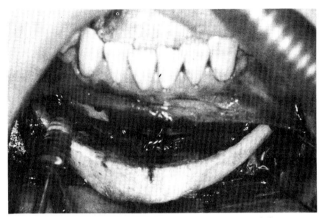

Fig. 11.15 Evening up by burring. The sharp edges and the posterior extremities are burred down in order to obtain à lower border which is regular and which is without palpable irregularities.

tensive posterior undermining, are easily adapted to the new position of the chin.

The muscular layer is sutured if possible, followed by suture of the mucosa, and a dressing is applied using elastic adhesive.

The suprabasilar osteotomy makes correction of different types of anomalies of the chin possible according to the direction of the osteotomy line, the use of an interpositional bone graft or, on the

other hand, the need to resect an intermediate segment, combined with an advancement of the mental fragment.

Thus the anterior sliding can be:

1. parallel to the occlusal plane (Fig. 11.16);
2. directed slightly inferiorly if one wants to increase the height of the chin;
3. associated with an elevation of the mental

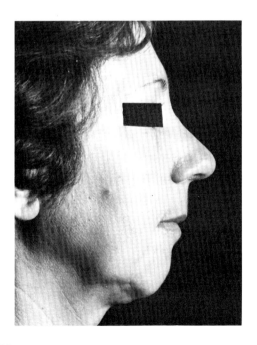

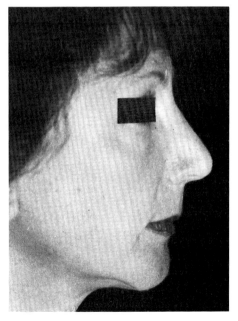

Fig. 11.16

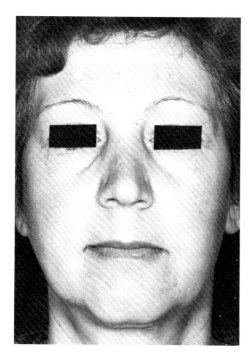 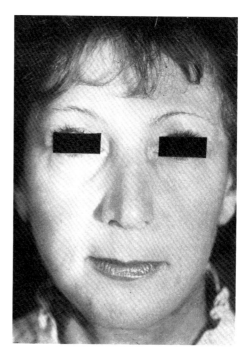

Fig. 11.16   Mental osteotomy with elevation of the inferior fragment riding up over the mandible (jumping genioplasty) with a 10 mm advancement. (The improvement in submental tension obtained by the osteotomy is completed by a face lift.)

fragment, which also permits, in cases of vertical excess, reduction of the height. In these cases, the mental fragment should ride over the body of the mandible for 2 to 4 mm; the perfect adaptation of the bony curves in contact is provided by using a bone burr.

An intermediate resection of a bony segment is indicated in rare cases of excessive height of the chin area.

Finally, an insufficiency in the height of the chin can be corrected by an intermediary bone graft (of iliac origin) (Fig. 11.17).

Conclusion

The suprabasilar osteotomy constitutes the treatment of choice for mental atrophies of moderate to severe degree, particularly when associated with anomalies in the height of the lower portion of the face, asymmetry or difficulties of labiomental function. However, silicone implants or insertion of the resected fragments from rhinoplasty have their indications in more discrete atrophies without functional anomalies.

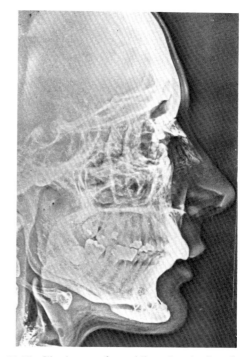

Fig. 11.17   Iliac bone graft providing a lengthening of 8–10 mm associated with an advancement of 9 mm.

BIBLIOGRAPHY

Aiach G, Levignac J 1986 La Rhinoplastie Esthétique. Masson, Paris, p. 176

Aiach G 1971 Traitement chirurgical des atrophies mentonnières. Revue de Stomatologie 72:(3): 439–448

Bell W H, Proffit W P, White R P 1980 Surgical correction of dento-facial deformities. Vol. II. W. B. Saunders, Philadelphia

Crestinu J M 1983 Periosteal incisions for mentoplasty. Plastic Reconstructive Surgery, 71(3) 440

Delaire J 1978 L'analyse architecturale et structurale cranio-faciale (de profil). Revue Stomatologie Paris 1: 1–33

Jack A, Friedland M D, Peter J, Coccaro D D S, Marquis J, Converse M D 1976 Retrospective cephalometric analysis of mandibular bone absorption under silicone rubber chin implants. Plastic and Reconstructive Surgery 57(2): 145

Mahler D 1982 Chin augmentation. A retrospective study. Annals of Plastic Surgery 8(5): 468

Tessier P L 1981 Chin advancement as an aid in correction of deformities of the mental and submental regions. Plastic and Reconstructive Surgery 67(5): 630

Tulasne J F, Raulo Y 1981 Excès vertical antérieur de l'étage inférieur de la face et génioplastie. Annales de Chirurgie Plastique 26(4) 332–336

Wolfe S A 1981 Chin advancement as an aid in correction of deformities of the mental and submental regions. Plastic and Reconstructive Surgery 67(5): 624

12. Reconstruction of the chin

A. Pasturel   J.-L. Cariou   P. Oxeda   P.-A. Diner   J.-M. Vallant

INTRODUCTION

The projecting portion of the mandible and the lower third of the face, the mental region is composed of several superimposed elements starting superficially and going deeper:

1. the skin — thick and hair-bearing in the male, covers a thin fatty layer crisscrossed by muscular fibres inserting into the skin;
2. the skin muscles — the quadratus of the chin, the muscles of the crest, the superior portion of the platysma of the neck, the inferior labial fibres of the labial orbicularis superiorly, and laterally the triangular muscles of the lips;
3. a skeletal structure — the mandibular symphysis supporting the inferior incisive segment onto the posterior surface of which insert the genioglossus and geniohyoid muscles as well as the anterior belly of the digastric;
4. the mucosal lining of the inferior gingivolabial cul-de-sac.

The chin is topographically separated from the lower lip by the labiomental crease, but very often this is involved and it is the entire labiomental region which must be corrected.

The role of this region is twofold:

Functional aspects. It participates in the maintenance of the continence of the oral cavity. It constitutes a point of principal insertion for the depressor muscles as well as the anterior attachments of the tongue and the hyoid bone (the pharyngo-linguo-hyo-mandibular complex).

Morphological aspects. It constitutes one of the principal facial projections which act as 'shock absorbers' in trauma. In profile, it counterbalances the projection of the nasal pyramid and participates in the vertical dimension of the face.

CLINICAL SITUATIONS REQUIRING RECONSTRUCTION

1. Trauma

Whether arising from road vehicle accidents, sporting accidents, accidents at work, accidental gun shot injuries or suicide attempts, lesions can be limited (wounds, scars) or more extensive (loss of substance).

One can group post-traumatic lesions into three large categories:

a. those only involving the skin;
b. those involving both the skin and the mucosa, exposing the mandible;
c. those composite lesions involving the bone, the mucosa and the skin.

2. Excision of cancer of the skin

This most often involves epidermal carcinomas of mucosal origin (of the inferior labial mucosa, gingival or anterior buccal mucosa) and extends to the mandibular symphysis and/or to the superficial integument by direct extension. More rarely, the origin is in skin or bone.

These excisions very frequently involve a portion of the buccal mucosa and the chin. The lower lip can be completely or partially included in the tumour resection. The bony resection should be marginal wherever possible, but the imperatives of cancer treatment do not always permit this, particularly in the edentulous patient whose anterior dental arcade is practically reduced to the lower border. The sacrifice of bony continuity completely alters the problems of reconstruction. It should be noted that a bilateral neck dissection, either radical or functional, is often associated with the local excision.

The consequences of large cancer excisions involving the labiomental region are multiple:

a. the muscular disequilibrium is responsible for the lateral displacement of the mandibular stumps under the action of the mylohyoids and the upward traction under the influence of the elevator muscles;
b. the loss of the anterior lingual attachments leads to a glossoptosis with oropharyngeal obstruction;
c. the buccal cavity is no longer continent;
d. swallowing is interfered with because of the loss of the elevating movements of the larynx;
e. the lower third of the face loses all projection; this is the 'Andy Gump Look' of the American authors, resulting in a 'Vögelgesicht' or 'bird face' profile.

PRINCIPLES OF RECONSTRUCTION

Whether dealing with trauma or cancer excision, even if the requirements are different, the aims of reconstruction are always functional and morphological, varying with the extent of the loss of substance.

Every reconstruction should aim to:

1. close the buccal cavity when it is open anteriorly,

2. reconstruct the labial muscular sling when it has been removed;

3. re-establish osseous mandibular continuity, permitting an eventual dental prosthesis;

4. obtain the most satisfactory morphological appearance.

TECHNIQUES OF RECONSTRUCTION

We will look at these as functions of the 'volume' of the loss of substance in its depth and surface extent.

Reconstruction of the cutaneous cover

This is most often required due to retractile scars, to losses of substance (purely cutaneous) from traumatic origins, to burn injury or to iatrogenic causes following removal of skin tumours.

Depending on the extent of the loss of substance, cutaneous cover can be provided by:

1. A full thickness skin graft, without forgetting that the skin in this region is the thickest of the facial area.
2. Local flaps, including the classic types of V–Y lengthening, Z plasty and advancement or rotation flaps.
3. Regional flaps. In this region they are numerous and we will cite the following:

   a. the jugal U-shaped flap of Mouly;
   b. the bipedical cervical flap;
   c. the jugo-cervico-supraclavicular flap;
   d. the frontal bipedicle flap of Kilner in women, carrying glabrous skin;
   e. the cranial bipedicle flap of Dufourmentel, brought down as a visor and used in the male because it is hair-bearing.

4. Distant flaps, of the delto-pectoral type or the distant tube pedical, are often more than is needed in these cases.
5. Cutaneous free flaps are rarely indicated in this group of lesions. Claude Le Quang has proposed, in cases of extended loss of genio-mandibular-mental skin in males, the use of a suprapubic flap, partially hair-bearing, taken on the two lateral iliac arteries.

Reconstruction of the loss of skin and mucosal substance exposing the mandible

This can be corrected by a single procedure in one operative session or by two different procedures in one or two sessions.

1. If two separate procedures are chosen:

   a. The mucosal lining can be reconstituted by cheek mucosal flaps, lateral genial cutaneous skin flaps turned over on themselves, or a thoracic cutaneous skin flap as described by Ginestet.
   b. Skin cover can be provided by Kilner's

bipedicle frontal flap (in the female) or by a bipedicle scalp flap of the Dufourmentel or Ginestet type (in the male), or by a musculocutaneous platysmal flap (Barron–Tessier).

2. The same surgical procedure can provide a monobloc cutaneous–mucosal reconstruction, and there are a number of possibilities:

a. Local and regional flaps. An example would be the association of two composite naso-genial flaps with inferior pedicles which can be brought together or superimposed along the midline and then associated with two cervical flaps.

b. Distant flaps. We will cite in this case the single or double delto-pectoral flap, providing a 'pharoah' type reconstruction as described by Soussaline. This greatly viable technique has the inconvenience of two operative stages and considerable scarring in the donor areas.

One should mention the possibility of a migratory abdominal tube pedicle, thoracic pedicle or brachial pedicle in multiple operative sessions.

c. Musculo–cutaneous flaps have, since their appearance in the therapeutic arsenal, considerably modified the possibilities of reconstruction of large losses of substance.

They have the advantage of:

(i) bringing two tissues into a monobloc unit: muscular for filling in, and skin of certain vitality for cover;

(ii) permitting primary closure of the donor area;

(iii) borrowing tissue which is not available locally from a distance.

One can criticise them for:

(i) their mutilating character;

(ii) the loss of muscular function caused by them;

(iii) the atrophy which occurs fairly rapidly of the transposed muscle, with the disappearance of mass in the long term

(which is not always a morphological inconvenience), and the appearance of a fibrous or retractile cord along the length of the tunnel created between the donor site and the recipient area;

(iv) their poor aesthetic appearance on the face due to their thickness, their texture, the poor colour match of the skin and particularly their immobility during facial movement;

(v) the difficulty, if not the impossibility of using a secondary dental prosthesis over this type of flap.

As regards the reconstruction of the labiomental region, great potential for myo-cutaneous flaps has been offered by certain muscles of the thorax (the pectoralis major, the latissimus dorsi), of the shoulder (the trapezius) and of the neck (the platysma, the sterno-cleidomastoid).

The pectoralis major myo-cutaneous flap, described by Ariyan in 1979, is the most widely used. Vascularized by the pectoral branches of the thoracoacromial artery, it has the advantages of great viability, of proximity, of rapidity of dissection and, finally of 'plasticity' in the design of its skin paddle (ovular with a greater vertical, oblique or horizontal axis, in a semilunar shape). Its indication is limited in women because of aesthetic considerations.

The latissimus dorsi musculo–cutaneous flap, initially used in 1906 by Tansini, has been widely used after the work of Olivari (1976) and McGraw (1979) and now rivals the pectoralis major flap. Vascularized by the thoracodorsal artery, a branch of the inferior scapular artery, it is taken as a vascular island on its pedicle and reaches the face by a transpectoral route or a subcutaneous route by tunnelling under the cervical skin. The dissection takes a little longer than that for the pectoralis major flap, but can often be done in the lateral decubitus position (J. M. Servant). The flap has numerous advantages: the dimensions of the skin paddle, its glabrous character, and the fact that the donor scar can be hidden.

The trapezius musculo–cutaneous flap, at present, is used in its lateral form with vascularization by the superficial branch of the transverse cervical artery described by Demergasso (1977), rather than in its inferior form described by Mathes and Nahai (1981), and pedicled on the deep branch. The blood supply can be compromised by a ballistic injury, or irradiation: this limits its indication, as does the difficulty of harvesting it.

The flap of the cervical platysma, described by Barron (1965) and vascularized by the submental artery, can be used in trauma cases for small losses of substance, but becomes risky if used in cancer patients. The same holds for the sternocleidomastoid musculo–cutaneous flap.

We should mention several technical points concerning the use of these flaps. During their harvesting, only the minimal muscular mass sufficient to provide nourishment and transport the skin paddle should be taken. In labiomental reconstruction, the paddle can be folded over like the pages of a book, its proximal portion reconstituting the mucosal intra-oral layer and the distal skin layer. The zone of folding thus corresponds to a new free labial border and is de-epithelialized in cases with persistent labial stumps which attach there. If possible, it is desirable to suspend the flap superiorly since the free border has a tendency to be pulled downward by the weight of the flap and the pedicle. Secondly, the pedicle can be sectioned, a myectomy performed, or the skin paddle partially defatted, in order to obtain a better blending of the contours with adjacent tissues.

    d. Free skin flaps, provided by microanastomosis and requiring the presence of recipient vessels of good quality, both arterial (facial artery, lingual artery, superior thyroid artery) and venous (external or internal jugular vein), are, at present, more often skin transfers of thin tissue, eventually resensitized: the 'Chinese' cutaneous antibrachial flap (Song 1978, Yang Guofan, 1981), the dorsalis pedis flap and the medial branchial flap. Alternatively, they may be partially hair-bearing transfers in men: the suprapubic flap already mentioned, or tranfers from other cutaneous areas which are thicker, such as the axillary flap, the scapular flap, the parascapular flap, the external mammary flap, the inguinal flap and so forth.

Reconstruction of composite osteo-cutaneo-mucosal losses of substance

It is here that one encounters all the imperatives of reconstruction and the methods used are numerous; they vary according to whether one decides to carry out the entire reconstruction in one or several stages.

Classically, one first reconstructs the covering layers and then the bony chin with a corticocancellous iliac graft in preference to a rib. The principal inconvenience of these bone grafts, the success rate of which is in the order of 80% (Salyer, Kruger) when they are carried out under good conditions, is related to their lack of proper blood supply leading to eventual resorption and disappearance, secondary local infection or poor vascular quality of the recipient bed. Thus, following the experimental work of Medgyesi (1973) and the publications of Snyder and Conley (1972) the notion has arisen of the necessity of using vascularized bone provided by composite osteo–myocutaneous flaps. At the same time, the development and the perfecting of microsurgical techniques has permitted the free transfer of composite grafts or pure osteoperiosteal segments.

The classical reconstruction in two stages

In a first stage, reconstruction of the skin cover and mucosal lining is carried out by one of the procedures that we have already mentioned: by the use of local flaps, local regional composite flaps, distant flaps or musculo-cutaneous flaps.

Here a maxillofacial prosthesis can be of great help. In effect, the placement of an endoprosthesis, maintaining the interfragmentary separation between the two bony stumps, makes it possible to control the scarring of the skin and acts as a conformer. This role of space maintenance can also be carried out by Kirshner wires or an external fixator.

At a second stage, an anterior mandibular arch is reconstructed with an autogenous corticocancellous iliac bone graft or a rib graft.

The iliac bone was first used for mandibular reconstruction in 1915. It provides a solid bony fragment, rich in cancellous bone, taken from either the mono- or bicortical iliac crest. On the other hand, adequate bending of the graft is often less easy to obtain than with a rib.

In order to avoid the harvesting of autogenous ribs, decalcified homografts have been used (Obson, Kaban), as well as lyophilized grafts (Sailer), irradiated grafts (Hamaker) or grafts treated by cryotherapy (Marx, Dougherty).

To recreate a satisfactory morphology of the mental region, it is necessary to obtain a curve of the graft which reproduces the loss of substance, and thus in some cases it will be preferable to take an iliac graft from the inner table rather than the outer table. In any case, the curve can be obtained either by small osteotomies carried out on the inner table (Millard) or by multiple parallel grooves on the posterior surface of a rib.

Let us recall some indispensable rules that give the best chances of success for a bone graft:

1. the necessity of a good recipient bed (scarred and irradiated tissues are poor indications);
2. the provision of an intimate contact between the cancellous surfaces of a graft and the mandibular segments (with an osteosynthesis in the tenon-mortise or swallowtail form);
3. the perfect immobilization of the graft by osteosynthesis (miniaturized miniplates, transosseous stainless steel wires) or a tray to carry the graft;
4. the maintenance of intermaxillary fixation for 6 to 8 weeks.

In cases of insufficient immobilization or of immobilization not being maintained for a long enough period, there is the risk of a pseudarthrosis developing; resorption of the graft can also come about because of partial exposure.

As an example we cite here the multi-stage reconstruction described by Ginestet: the provision of mucosal lining by a tube pedicle flap with a thoracic paddle, of skin cover by a bipedicle scalp flap, with an iliac bone graft at a later date.

One stage reconstruction

This requires composite flaps containing bone, pedicled or free. The notion, which is not recent, having been published by Israel in 1896, inspired by the brachial flap of Tagliacozzi (1597), transposes a segment of the ulna pedicled on its skin for cover in the reconstruction of the nose. In 1918, Blair inaugurated the technique of reconstruction of losses of mandibular substance by transferring a segment of clavicle pedicled on a vertical skin flap. In 1972, Conley reported a number of local composite flaps (anterior or posterior cervicle, and thoracic) which can be used with this type of indication. However, these flaps see their role limited since the bony segments are in general of small length and of mediocre vitality. It is, in fact, the arrival of microsurgical techniques in clinical practice on the one hand, and the progress achieved in the techniques of reconstruction of the covering layers by the introduction of musculo-cutaneous flaps on the other, that has made possible the transfer of living osseous segments to the face. It was Cuono and Ariyan who, in 1979, synthesized these two notions: they carried out the first one stage reconstruction of an osteo-cutaneous loss of substance in the lower level of the face by using an osteo-myocutaneous pectoralis major flap carrying the fifth rib.

Thus composite flaps were born, pedicled or free, carrying vascularized bone which is capable of integrating with local structures as well as resisting infection and the phenomenon of resorption.

These composite free flaps were the first to be utilized. Numerous donor sites have been described, of which can be mentioned:

1. *The iliac flap*, initially described as an osteo-cutaneous flap by Taylor in 1975.

At present this flap, leaving minimal sequelae both in aesthetic and functional areas, makes extensive reconstructions possible and may be taken in three different manners:

   a. as a pure iliac graft, isolated on the deep circumflex iliac vessels, and requiring the harvesting of a certain amount of the transversalis and external and internal oblique muscles;
   b. as a vascularized osteo-cutaneous flap using the deep circumflex iliac pedicle (Taylor 1980);
   c. as an osteo-cutaneous flap with a larger skin paddle taken on two pedicles —

the superficial circumflex iliac and the deep circumflex iliac (Salibian 1986).

2. *The dorsalis pedis flap* carrying the second metatarsal, used by O'Brien (1979) and Rosen (1979), is of limited length (< 7 cm) and is not very malleable.

3. *Rib flaps*, anterior or posterior, from the 9th or 10th rib, have been proposed by numerous authors, and can be:

  a. bony: Doi (1976), Buncke (1977), Daniel (1977), Harashina (1978);
  b. osteo-cutaneous: Ariyan (1978), Daniel (1979), Serafin (1980).

4. *The scapular osteo-cutaneous flap* (Swartz 1986), an extension of the parascapular flap (Nassif 1982) or of the scapular flap (Dos Santos 1984), and vascularized by the circumflex scapular artery, is of the same viability as the iliac crest flap.

Compared to the latter it has the advantage of providing a finer segment of bone which is purely cortical.

5. *The 'Chinese' antibrachial radial flap* (Song 1978, Yang Guofan 1981) was used as an osteo-cutaneous flap by Biemer in 1983 and then Soutar for mandibular reconstruction. If the thin skin paddle is a good replacement for buccal mucosa and permits an eventual dental prosthesis, harvesting has two inconveniences: the aesthetic sequel and the length of the straight segment of bone being limited to 10 cm.

All of these flaps, revascularized after microanastomosis, have the advantage of providing bone vascularized by a medullary network or by a musculo-periosteal network depending on the method of harvest. Here a controversy exists because, for certain authors the periosteal vascularization (which is the case for the anterior costal arch) is sufficient for the viability of the bone transplant, but for other authors the only useful sites are those which preserve both the periosteal and the medullary blood supply (which is the case with the iliac crest).

The two donor sites most often used at present are without doubt the iliac and antebrachial.

Besides the need for harvesting bone, these revascularization transfers require a long operation, experience with microsurgical techniques, and recipient vessels of good quality, without forgetting their risk of failure due to thrombosis of the anastomosis.

More recently, osteo-cutaneous flaps carried by a muscular pedicle have come to complete the therapeutic arsenal.

Since Cuono, who in 1980 carried out a mandibular reconstruction using an osteo-myocutaneous pectoralis major flap with a fifth rib, other composite flaps usable as pedicled or 'island' flaps have been described.

These are:

1. *The osteo-myocutaneous flap of the pectoralis major* with the fifth or sixth rib according to the authors, or the fourth rib (making viability more certain according to Little);

2. *The osteo-myocutaneous flap of the latissimus dorsi* (Panconi 1981) carrying the ninth or tenth rib; this presents a better viability of its osseous component and leaves fewer sequelae;

3. *The trapezius osteo-myocutaneous flap* (Demergasso 1979), which has the advantage of thinner bone, carrying the acromion or the scapular spine;

4. *The sternocleidomastoid osteo-myocutaneous flap* with a segment of clavicle, which does not seem to have sufficient viability, and certain local vascular conditions may render it unusable.

When compared to the microsurgical techniques, the osteo-myocutaneous flaps have the advantage of being more easily harvested and the absence of an anastomosis, implying a shorter operation as well as a better viability. But one must remember that the vitality of their bony component is controversial because here there is only periosteal vascularization, and they are less malleable with their osseous and skin components.

INDICATIONS

Discussion remains open at present about the chronology of the immediate or secondary reconstruction, in one or several stages. Before providing the indications, we should take count of certain general and local factors:

— generally, in particular the age (in the older subject we prefer classic procedures which require a short operation) and the physiological burdens;

— and locally, the 'volume' of the loss of substance, the amount of scar tissue in the region of the defect, the vascular quality of the recipient site, associated lesions and other observations made during the course of the operation.

All of these elements vary depending on whether the loss of labiomental substance is due to trauma or excision of cancer.

1. **Post-traumatic defects.** The indication is essentially a function of the age of the patient, the sex of the patient (hair-bearing areas in the male), the surgical experience of the operator, and the extent of the loss of substance both in surface area and depth.

a. In the first group of lesions — those which are purely cutaneous, one generally uses either a skin graft or a local flap, respecting as much as possible the principle of replacing like tissue with like.

b. The reconstruction of muco-cutaneous lesions can be carried out in one or several stages according to different authors. We would like to recall the usefulness of the association of two nasogenial flaps with inferior pedicles (Fig. 12.1).

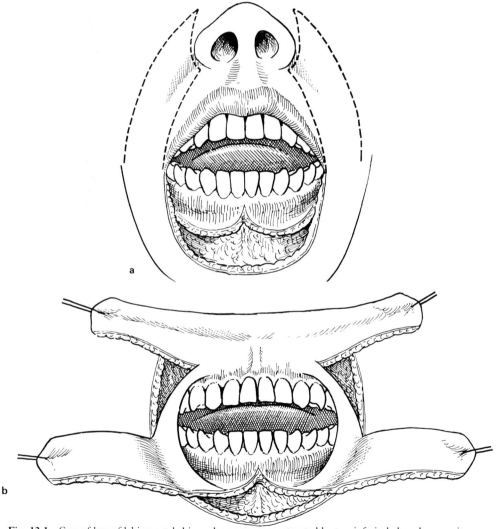

Fig. 12.1 Case of loss of labiomental skin and mucosa, reconstructed by two inferiorly based composite nasogenial flaps. **a.** Design of the flaps; **b.** elevation of the two 'vascularized' flaps which are going to be rotated by 90 degrees.

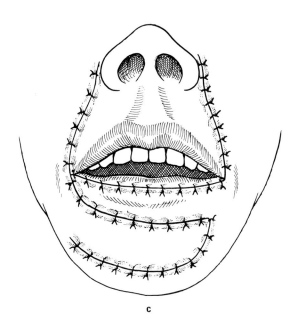

c

Fig. 12.1 c. The transfer of the naso-genial flaps to reconstruct the loss of substance, the muco-cutaneous line of junction of the lower lip being reconstituted by bringing the mucosa over the upper border of the superior flap.

c. The third group, osteo-cutaneo-mucosal lesions, includes the post-traumatic loss of substance in the labiomental region due to ballistic injuries.

After initial treatment (cleaning of the wound, debridement only when essential, and maintenance of spaces by means of an external fixator or an endoprosthesis), reconstruction is most often carried out in two stages.

The *first stage*, involving skin coverage and mucosal lining, calls for a musculo-cutaneous flap (the pectoralis major or latissimus dorsi) where the skin paddle is turned like a book flap;this is preferable to a delto-pectoral flap in the shape of a 'camel's hump' which requires a secondary division and provides a poorer quality recipient bed for the eventual bone graft;

A *second stage* involves the re-establishment of mandibular osseous continuity by a corticocancellous iliac or costal graft, introduced by an extra oral approach.

If reconstruction in one stage is chosen, a free osteo-cutaneous flap (iliac, scapular or radial) is preferred to an osteo-myocutaneous flap where the bony segment has less vitality. In cases where microsurgery is decided upon (Fig. 12.2), it is necessary to establish the quality of the cervical recipient vessels as they can be the site of macro- or microscopic lesions caused by the cavitation phenomenon of the original ballistic injury, which can cause early or late thrombosis of microanastomoses. In cases where there is doubt, it is better to perform the arterial and venous anastomoses at a distance and bridge the gaps with saphenous vein grafts.

2. **Reconstruction of defects following cancer excision.**

The poor prognosis of epitheliomas of the anterior buccal cavity (20–25% 5 year survival) and the difficulties in reconstruction have led to the adoption of an approach delaying osseous reconstruction or even dispensing with it completely.

a. The restoration of the soft tissues is always carried out immediately to assure the continuity of the buccal cavity.

If the defect following excision is confined to the labiomental region, two nasogenial flaps or a double pectoral flap can be used.

If the size of the defect is deeper, or involves the floor of the mouth, one can choose either an osteo-cutaneous flap (pectoralis major, trapezius or latissimus dorsi), or the association, for example, of a pectoralis major musculo-cutaneous flap and a delto-pectoral flap (Fig. 12.3).

It is in these cases that certain authors have proposed an immediate composite reconstruction using an osteo-myocutaneous flap (pectoralis major, trapezius, latissimus dorsi) on a pedicle or an osteo-cutaneous free flap (iliac, scapular). In spite of the advantages provided by revascularized bone grafts, they are often indicated only in selected patients who are in good general health, where the magnitude of the tumour excision is not great and where no radiotherapy is envisaged.

b. In general, osseous reconstruction is deferred for a varying amount of time to provide a period of observation for recurrent cancer.

This approach is most justified when the patient is in poor overall condition, the extent of cancer

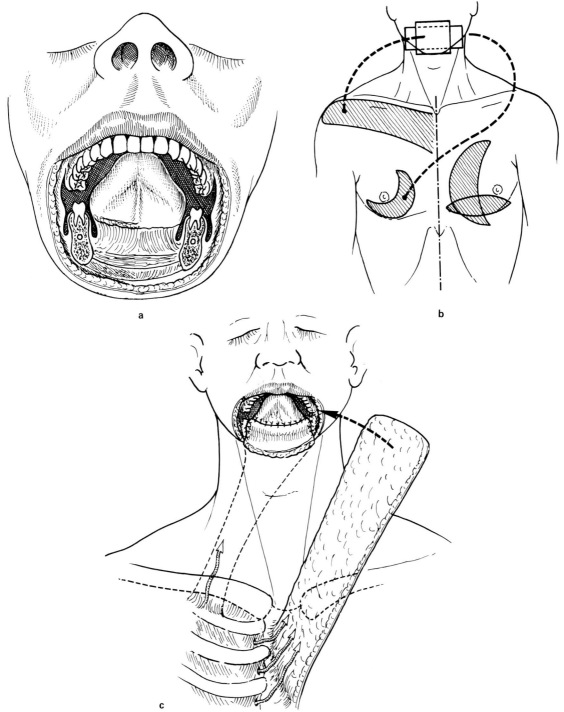

Fig. 12.2 Case of loss of all three types of tissue of the labiomental area extending to the anterior floor of the mouth with the removal of the anterior arc of the mandible, where the first reconstructive procedure will be only of soft tissues. **a.** Loss of substance; **b.** design of the cutaneous paddles of the deltopectoral and pectoralis major musculo-cutaneous flaps (with their vertical, horizontal and semi-lunar variations); **c.** the skin paddle of the pectoralis major musculo-cutaneous flap (tunnelled in the cervical region) reconstitutes the anterior floor of the mouth and the median gingivolabial covering of the future alveolar ridge.

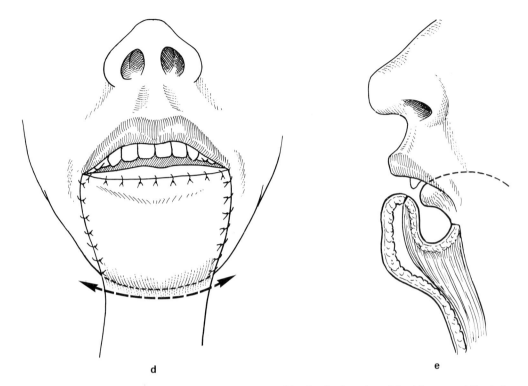

d

e

Fig. 12.2 d. The cutaneous lip and the chin are reconstructed by the distal portion of the deltopectoral flap (tube pedicle) which will be divided on the fifteenth day; **e.** profile view of the facial reconstruction.

invasion is great, and where postoperative radiotherapy will be necessary.

During the waiting period it seems desirable to maintain the interfragmentary separation either by Kirshner rods or better still by a plastic or metallic endoprosthesis. The latter, upon which the base of the tongue and the suprahyoid muscles are reinserted, is fixed to the mandibular stumps and limits the retraction of the reconstructed soft tissues.

At the time of the bone graft, it is necessary to evaluate the quality of the recipient bed carefully. In certain favourable cases, one can use a free iliac graft if the skin and lining cover permit and if the loss of bony substance is moderate: this should be placed by a cutaneous approach. More often, the extent of scarring or retraction of the soft tissues which have 'weathered the storm' of radiotherapy, as well as the need to bring in more soft tissue,

Fig. 12.3 Reconstruction of loss of three types of tissue in one stage by a free osteo-myocutaneous iliac flap attached to the cervical vessels. **a.** At the level of the inguinal donor site, the two pedicles of the superficial circumflex iliac artery (the skin flap) and the deep circumflex iliac artery (the myo-osseous and the osteo-myocutaneous flaps) can be used together or separately; **b.** schematic view of the harvesting of the osteo-myocutaneous flap based on these two pedicles; **c.** the flap is transferred to the lower portion of the face. The circumflex iliac arteries are anastomosed — the deep to the facial artery, and the superficial to the superior thyroid artery. The venous anastomoses are carried out at the confluence of the thyro-linguo-facial veins; **d.** view in profile showing the reconstruction of the anterior arch of the mandible with the iliac crest osteosynthesized by miniature plates screwed into place, and the linguo-labio-mental reconstruction with the inguinal skin paddle.

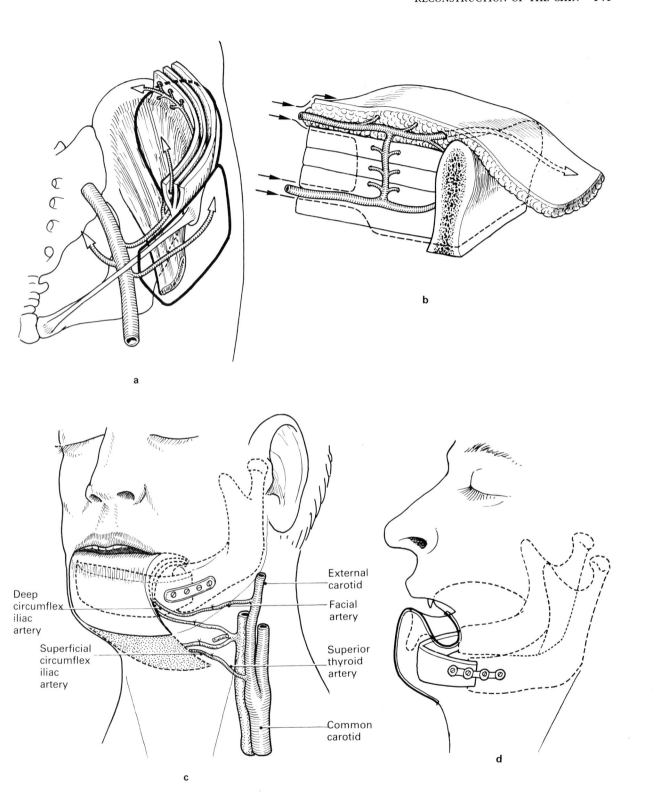

a

b

Deep
circumflex
iliac
artery

Superficial
circumflex
iliac
artery

External
carotid

Facial
artery

Superior
thyroid
artery

Common
carotid

c

d

lead one to opt for vascularized bone in the form of a microsurgical bone transfer of a composite osteo-cutaneous flap.

CONCLUSION

Reconstruction of the labiomental region, whether it involves the anterior arch of the mandible or not, should restore the best possible function and morphology to the patient by assuring the continuity of the buccal cavity.

Without rendering classical local flaps and bone grafts obsolete, two more recent procedures in the therapeutic arsenal have considerably modified the possibilities of reconstruction: the musculo-cutaneous flaps on the one hand and revascularized bone grafts or composite flaps on the other hand.

If aesthetic considerations constitute one of the major preoccupations in the choice of a procedure to 'resurface' post-traumatic lesions of the first and second groups, other considerations become important in situations where there has been loss of substance due to gun shot wounds or in the reconstruction of defects following cancer surgery, where the aim is to restore the patient as well as possible while at the same time respecting the imperatives of cancer treatment.

The choice of technique employed and the chronology must be discussed case by case; the drawbacks of techniques, which are often very sophisticated and have their advantages and disadvantages must be considered, as well as the benefits which the patient can draw from them, bearing in mind the extent of the mutilation or illness.

13. Long-term results of genioplasty

J. F. Tulasne   G. Despreaux

The genioplasty is an operation which remodels the chin. By a horizontal subapical osteotomy, the basilar portion of the mandibular symphysis is mobilized, remodelled with a burr and fixed in its new position by three osteosynthesis wires (Fig. 13.1). Our study only included anterior and superior displacements which are the most frequent, the so-called 'jumping bone flap' technique of Anglo-Saxon authors. The results of this technique have been analyzed by Dr. Despreaux in his doctoral thesis submitted in 1984 in which he studied 22 adult patients (Table 13.1). We have added to this the results seen in 17 infants between $9\frac{1}{2}$ and 15 years of age, who we have also operated upon and who have been followed radiologically for at least 5 months following surgical intervention (Table 13.2).

The methodology and analysis of results that are presented here are taken from Dr. Despreaux's thesis.

METHODOLOGY (Figs 13.2 and 13.3)

The landmark used for antero-posterior measurement is the posterior cortex of the mental symphysis, the superior portion of which is not involved in the genioplasty. This portion of the symphysis does move along with the body of the mandible in cases of extensive mandibular osteotomy with which a genioplasty is often associated. This makes possible the study of movements due to the genioplasty alone.

In practice, this has been carried out by calculating from the pre-operative radiograph and

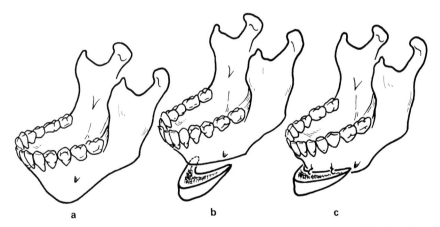

Fig. 13.1 Technique of the overriding genioplasty ('jumping bone flap'). **a.** After dissection of the symphysis through a vestibular incision, the horizontal suprabasilar osteotomy is carried out with a saw; **b.** the fragment is partially freed from its muscular and periosteal attachments. It is remodelled with a burr so as to fit perfectly on the anterior surface of the symphysis; **c.** stabilization is provided by three stainless steel osteosynthesis wires. The soft tissues are stretched to adapt to the new shape of the chin. The incision is closed in two layers: muscular and mucosal.

Table 13.1 Genioplasties in the adult

Case	Sex	Age	Diagnosis	Associated procedure	Length of followup
1	F	24	Isolated retrogenia	Rhinoplasty	6
2	F	23	Isolated retrogenia		7
3	M	17	Sequelae of cleft lip	Rhinoplasty	8
4	F	32	Anterior open bite	Köle	12
5	F	27	Bimaxillary protrusion	Wassmund + Köle	8
6	M	26	Vertical maxillary excess	Le Fort I	12
7	F	19	Anterior open bite	Le Fort I	12
8	F	17	Class III	Obwegeser	12
9	M	20	Class III	Obwegeser	10
10	M	17	Class III	Obwegeser	11
11	F	31	Class III	Obwegeser	6
12	F	18	Class III	Le Fort I	6
13	M	23	Class III	Le Fort I	10
14	M	24	Class III	Le Fort I	12
15	F	38	Class III + Anterior open bite	Köle	8
16	M	23	Class III + Anterior open bite	Le Fort I + Köle	11
17	F	20	Class III + Anterior open bite	Le Fort I + Obwegeser	12
18	F	24	Class II	Obwegeser	28
19	F	21	Class II + Maxillary crossbite	Le Fort I + Obwegeser	6
20	F	18	Micromandibular anomaly	Inverted L-osteotomy of the ascending rami	14
21	F	23	Micromandibular anomaly	Inverted L-osteotomy of the ascending rami	52
22	F	16	Micromandibular anomaly	Inverted L-osteotomy of the ascending rami	46

Table 13.2 Genioplasties in the child

Case	Sex	Age	Diagnosis	Associated procedure	Length of followup
1	M	9	Micromandibular anomaly	Inverted L-osteotomy of the ascending rami	36
2	F	11	Class II	Wassmund	11
3	F	11	Class II	Obwegeser	5
4	F	12	Class III	Obwegeser + glossectomy	6
5	M	12	Class II	Le Fort I	5
6	F	12	Class II	Obwegeser	16
7	M	13	Class II	Obwegeser	8
8	M	13	Class II	Obwegeser	5
9	F	13	Micromandibular anomaly	Inverted L-osteotomy of the ascending rami	17
10	M	13	Class II	Maxillary incisive–canine corticotomy	15
11	M	14	Anterior open bite	Glossectomy	11
12	F	14	Class II	Wassmund + inferior segmental incisive osteotomy	18
13	M	14	Mandibular asymmetry	Inverted L- osteotomy of the ascending ramus + sagittal split Obwegeser	5
14	M	14	Class II + crossbite	Le Fort I + Obwegeser	14
15	M	14	Class II + vertical maxillary excess	Le Fort I + Obwegeser	6
16	M	15	Class II + crossbite	Le Fort I	19
17	F	15	Anterior open bite	Le Fort I	12

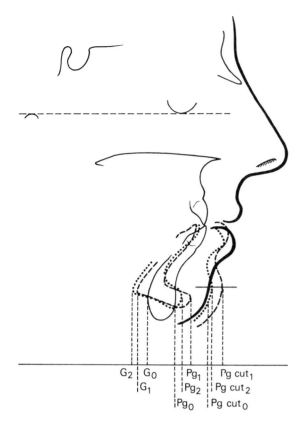

Fig. 13.2 Complete analysis in a case of mandibular setback with genioplasty. Superposition of three radiographs: pre-operative (solid line, reference 0); immediate postoperative (dashed line, reference 1); mean or long-term postoperative result (dotted line, reference 2).

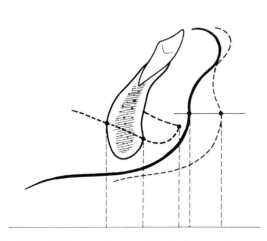

Fig. 13.3 Reference points used in the absence of movement of the mandibular body, with superposition of two radiographs taken before and soon after the genioplasty.

considering the osseous and dental elements (the cranial base, the maxilla and the mandible with the outline of the posterior cortex of the symphysis) as well as the cutaneous profile. Then the osseous and cutaneous displacements caused by the operation as well as the postoperative modifications were registered by superimposing the studies in the cranial base region.

The study of the early patients showed the need to define a fixed cortical point vertically because the posterior surface of the symphysis is neither rectilinear nor frontal: the point G_0 is the intersection, on the superimpositions, of the posterior limit of the cortical symphysis pre-operatively with the axis of the horizontal osteotomy. Only in a few cases, under unusual conditions, was a different construction of point G_0 used.

The subsequent calculations made it possible, according to the same criteria, to define points G_1 by a postoperative radiological examination, and then G_2 corresponding to the first film taken after a minimum period of 6 months, and eventually G_3 by the final radiological examination.

The different measurements were made by performing an orthogonal projection of elements to be studied on a parallel to the horizontal Frankfort plane and measuring the abscissae thus obtained from a zero reference point corresponding to the projection of point G_0 on this line.

The points Pg_0 and Pg Cut_0 correspond respectively to the most anterior points of the mental bony symphysis and the mental soft tissues pre-operatively.

In certain cases, due to the retrusive position of the deformed chin, the establishment of points Pg_0 and particularly Pg Cut_0 required an arbitrary construction (see Despreaux 1984).

Points Pg_1 and Pg Cut_1 represent respectively the most anterior points of the bony segment and of the cutaneous chin, observed on the first X-ray taken immediately after the operation. The points Pg_2 and Pg Cut_2 were obtained in the same manner on a film taken six months following surgery.

The measurements taken in the antero-posterior direction were as follows:

1. Intra-operative advancement of the bony chin measured by the difference between the projections of the distances G_1–Pg_1 and G_0–Pg_0;

2. Advancement of the bony chin at mid-term (a minimum of 5 months after surgery), measured by the difference between G_2-Pg_2 and G_0-Pg_0;
3. Advancement of the bony chin in the long-term when we had a final film at our disposal, measured by the difference between G_3-Pg_3 and G_0-Pg_0;
4. Advancement of the cutaneous chin in the medium and long-term.

These measurements made possible the determination of two elements which are of interest to the surgeon: the postoperative osseous resorption (expressed as a percentage) and the relationship between the cutaneous advancement at mid-term examination and the osseous advancement obtained at the time of operation.

RESULTS

1. In the adult (Table 13.3 and Figs 13.4–13.7):

a. The mean interoperative advancement of the bony chin was 9.9 mm with a range of 3 to 14 mm;
b. The mean advancement of the bony chin at mid-term was 6.9 mm with a range of 3 to 12.5 mm.
c. The mean advancement of the cutaneous chin at mid-term was 7 mm with a range of 3 to 13.5 mm.

At midterm evaluation:

1. There was a mean osseous resorption of 2.5 mm in 20 cases. This represents a mean of 25% of the intra-operative bony advancement, with a range of 0 to 89%, this value varying little if one excludes the 3 highest and lowest values of the series.
2. The remodelling of the bone is fairly typical, combining resorption and apposition (Fig. 13.8). Resorption is predominant along the antero-superior border of the bony segment, thus displacing the pogonion point (Pg) downwards between the

Table 13.3 Results of genioplasties in the adult

| Case | Intra-operative bony advancement (mm) | Mid-term (> 6 months) | | | |
		Bony advancement (mm)	Bony resorption (%)	Cutaneous advancement (mm)	Cutaneous advancement over interoperative bony advancement (%)
1	8.5	6	17	7	82
2	10.5	9	14	9	85
3	10.5	8.5	19	9	85
4	11	9	18	7	63
5	11	6	45	5	45
6	?	4.5	?	3	?
7	14	12	14	9	64
8	11	9.5	13	8.5	77
9	8.5	5	41	6	70
10	8	6	25	5	62
11	7	3.5	50	5	71
12	10	8.5	15	5.5	55
13	11.5	9	21	7	60
14	10.5	8.5	19	9	85
15	?	3.5	?	3	?
16	3	3	0	3	100
17	7.5	3.5	53	4	53
18	7.5	7	6	6	80
19	10	9	10	7	70
20	14	7	50	13.5	96
21	14	12.5	10	13.5	96
22	11.5	7	39	7	60
Mean	9.9	6.9	25	7	73

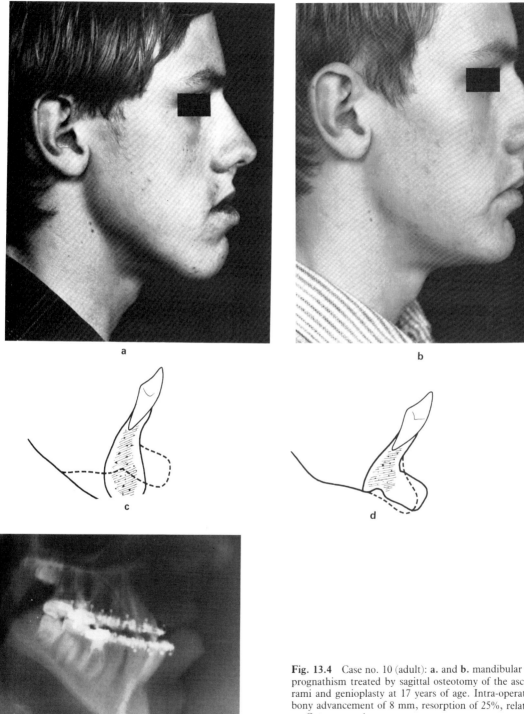

Fig. 13.4 Case no. 10 (adult): **a.** and **b.** mandibular prognathism treated by sagittal osteotomy of the ascending rami and genioplasty at 17 years of age. Intra-operative bony advancement of 8 mm, resorption of 25%, relationship of $\frac{\text{Cutaneous advancement}}{\text{Bony advancement}} = 62\%$.

c. Pre-operative + immediate postoperative.
d. Immediate postoperative + postoperative after 6 months.
e. Pre-operative.

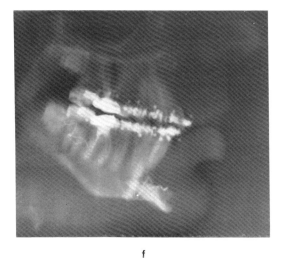

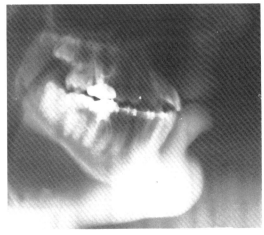

f

g

Fig. 13.4
f. Immediate postoperative.
g. Postoperative (minimum 6 months).

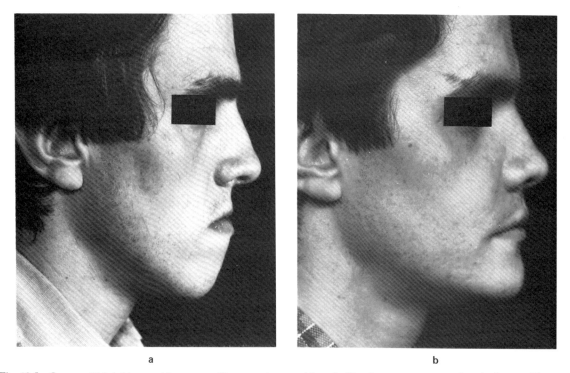

a

b

Fig. 13.5   Case no. 14 (adult): **a.** and **b.** retromaxillar anomaly treated by a Le Fort I type osteotomy and genioplasty at 24 years of age. Bony interoperative advancement of 10.5 mm, resorption of 19%, relationship of $\dfrac{\text{skin advancement}}{\text{bony advancement}} = 85\%$.

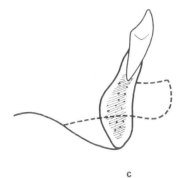

c

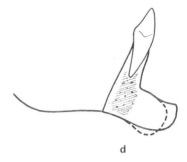

d

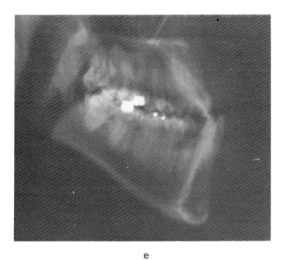

e

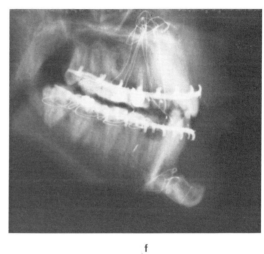

f

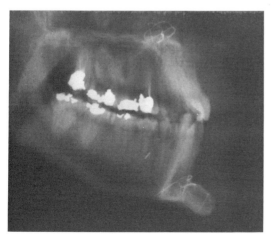

g

Fig. 13.5

c. Pre-operative + immediate postoperative.
d. Immediate postoperative + postoperative after 6 months.
e. Pre-operative.
f. Immediate postoperative.
g. Postoperative (minimum 6 months).

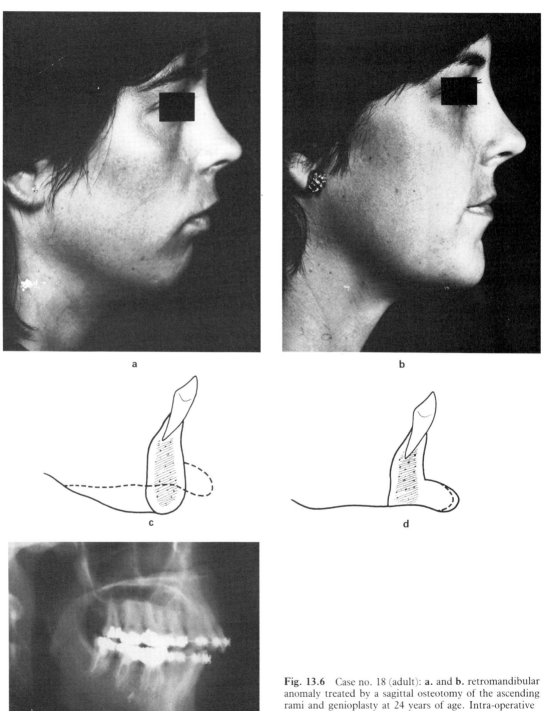

a

b

c

d

e

Fig. 13.6 Case no. 18 (adult): **a.** and **b.** retromandibular anomaly treated by a sagittal osteotomy of the ascending rami and genioplasty at 24 years of age. Intra-operative bony advancement of 7.5 mm, resorption of 6%, relationship of $\frac{\text{skin advancement}}{\text{bony advancement}} = 80\%$.

c. Pre-operative + immediate postoperative.
d. Immediate postoperative + postoperative after 6 months.
e. Pre-operative.

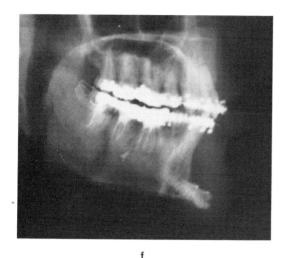

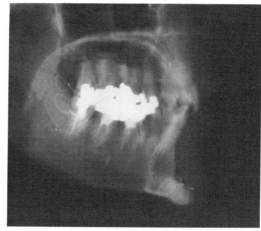

f

g

Fig. 13.6
f. Immediate postoperative.
g. Postoperative (minimum 6 months).

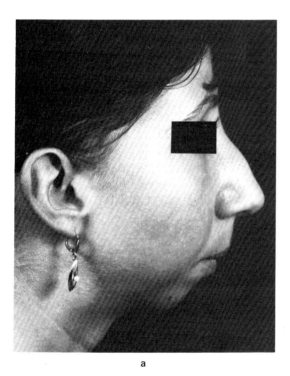

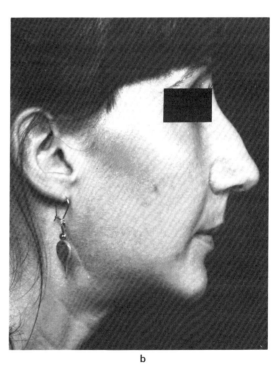

a

b

Fig. 13.7 Case no. 21 (adult): micromandibular deformity due to rheumatoid arthritis treated by inverted L-osteotomy of the ascending rami with an iliac bone graft and a genioplasty at 23 years of age. Intra-operative bony advancement of 14 mm, resorption of 10%; relationship of $\dfrac{\text{skin advancement}}{\text{bony advancement}} = 96\%$.

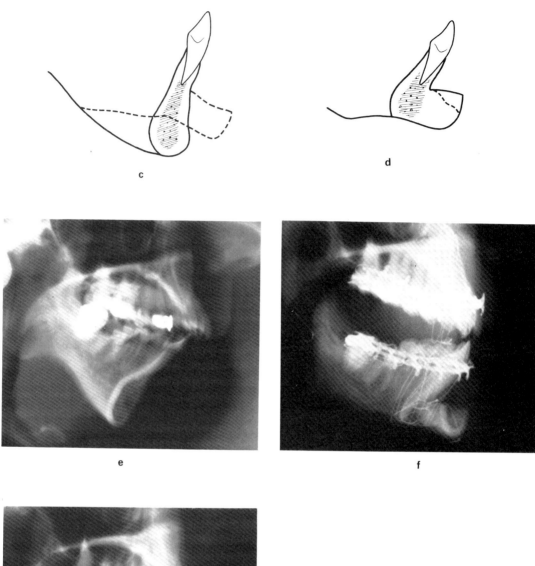

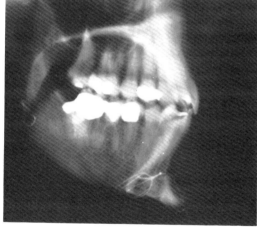

Fig. 13.7
c. Pre-operative + immediate postoperative.
d. Immediate postoperative + postoperative after 6 months.
e. Pre-operative.
f. Immediate postoperative.
g. Post-operative (minimum 6 months).

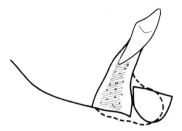

Fig. 13.8 Drawing showing the usual remodelling of the fragment with both resorption and bony apposition. Generally, these phenomena are much more marked in the child.

immediate postoperative film and the mid-term film. Furthermore, bony appositional growth seems to occur in the angle between the bony segment and the anterior surface of the symphysis, as well as along the section of the lower border of the mandible.

3. The association of a genioplasty with an advancement of the mandible does not seem to influence the percentage of resorption of the fragment: this in fact seems to be higher in our series of sagittal osteotomies with mandibular setback.

4. The relationship of cutaneous advancement with the intra-operative bony advancement is the most important element to be considered, since this enables predictions to be made and the surgical approach adjusted accordingly. In this study, the mean relationship was 73%, with a range of 45 to 100%.

On long-term evaluation the advancement of the chin seems to be stabilized by 6 months postoperatively. The 8 cases that we able to study after 6 months showed minimal or no resorption of the bony chin between the sixth and twelfth month.

2. **In the child** (Table 13.4 and Figs 13.9 and 13.10):

    a. The 17 cases studied were those of 10 boys and 7 girls between $9\frac{1}{2}$ and 15 years of age. The genioplasty was always associated with another osteotomy, except in case 11 (Table 13.2). The mean duration of observation was 12 months with a range of 5 to 36 months.

Table 13.4 Results of genioplasties in children

Case	Age	Interoperative bony advancement (mm)	Bony resorption at mid-term (>5 months) (%)
1	9	11	75
2	11	8.5	35
3	11	9	33
4	12	8	25
5	12	15	40
6	12	13	46
7	13	13	30
8	13	13	23
9	13	10	50
10	13	15	53
11	14	14.5	24
12	14	10	40
13	14	11	63
14	14	15	40
15	14	13	46
16	10	10	40
17	13	13	7
Mean		11.9	39.4

    b. The mean intra-operative advancement of the chin was 12 mm, with a range of 8 to 15 mm.

    c. The resorption of the fragment of the genioplasty is clearly greater than in the adult since it reached a mean of 40%. Only 4 cases showed a resorption in the range seen in adult cases, that is to say 25% or less.

    d. Three cases could be studied more thoroughly because of cephalometrical film taken in the second postoperative month. At this stage, the resorption was less than 10% of the advancement. Ultimately, the resorption went on to 33% in one case (at the fifth month), to 40% in another case (at the seventh and fourteenth month) and to 53% in the third case (at the fifteenth month).

    e. The remodelling of the lower border segment is particularly important, concerning resorption as much as the phenomenon of bony apposition.

    f. Finally, just as in the result, there seems to be practically no more resorption after the sixth postoperative month.

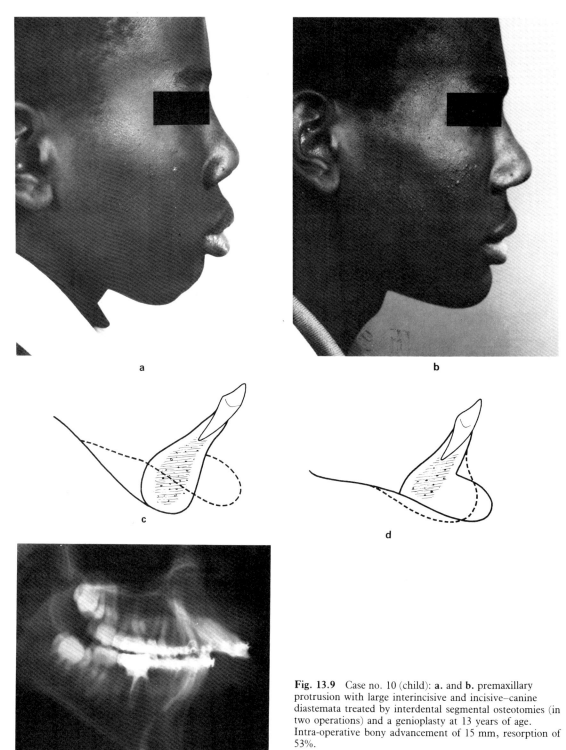

Fig. 13.9 Case no. 10 (child): **a.** and **b.** premaxillary protrusion with large interincisive and incisive–canine diastemata treated by interdental segmental osteotomies (in two operations) and a genioplasty at 13 years of age. Intra-operative bony advancement of 15 mm, resorption of 53%.

c. Pre-operative + immediate postoperative.

d. Immediate postoperative + postoperative after 6 months.

e. Pre-operative.

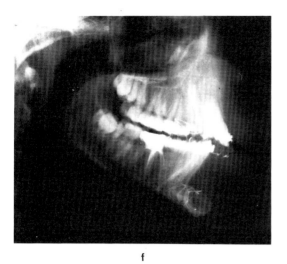
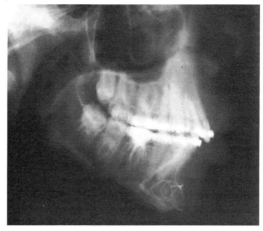

f   g

Fig. 13.9
f. Immediate postoperative.
g. Postoperative (minimum 6 months).

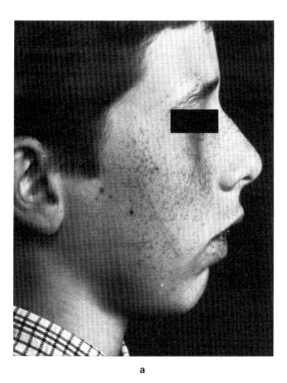
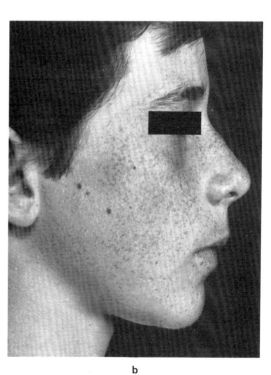

a   b

Fig. 13.10   Case no. 14 (child): Retromandibular anomaly and a vertical maxillary excess treated by a Le Fort I type osteotomy, a sagittal osteotomy of the ascending rami and a genioplasty at the age of 14. Bony interoperative advancement of 13 mm, resorption of 46%.

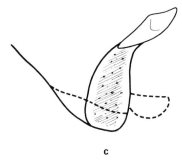 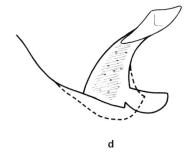

c   d

Fig. 13.10

CONCLUSION

The long term results of genioplasty performed by the 'jumping bone flap' show that it undergoes a resorption of the bony fragment reaching a mean of 25% of the advancement carried out in the adult and 40% in the child. These results are stable after the sixth month, most of the resorption having taken place after the third postoperative month. The study of the soft tissues shows that in the adult the cutaneous chin advancement is 75% that of the advancement of the bony chin.

BIBLIOGRAPHY

Delaire J, Tulasne J F 1979 Déséquilibres labiomentonniers par excès vertical antérieur de l'étage inférieur de la face. Apport de la géniectomie segmentaire horizontale. Orthodontie Francaise 50(2): 353–375

Despreaux G 1984 La génioplastie par chevauchement dans le syndrome d'excès vertical antérieur de la face. Thèse de Doctorat en Médecine, Faculté de Médecine de Bobigny, Université Paris Nord

Tulasne J F, Raulo Y 1981 Excès vertical antérieur de l'étage inférieur de la face et génioplastie. Annales de Chirurgie Plastique 26(4): 332–336

14. The chin in the short face

J. Dautrey

In the short face, the lower portion or the maxillomandibular portion is reduced in relation to the middle portion. The insufficient height is associated with inadequate development of the anterior regions of the maxilla and/or the mandible.

The most classic case is the vertical underdevelopment of the symphyseal region of the mandible.

The typical facial appearance of these patients includes: a folding or rolling of the lower lip outwards causing an unattractive labiomental fold; vertical underdevelopment of the mental region producing a concave inferior border (which is nor-

mally convex), sometimes causing a vertical dimple.

On the other hand, the profile of the chin can, in fact, project, thus accentuating the labiomental fold.

If one examines the alveolar process, one often notes an elevation in this area causing a supra-eruption of the inferior incisors.

The free border of the inferior incisors can thus come into contact with the palatal mucosa, occasionally causing an ulceration. The photographs, cephalometric films of the face and panoramic X-rays (Figs 14.1 and 14.2) taken before the

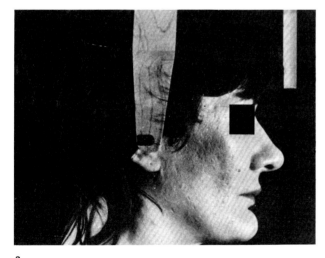

a

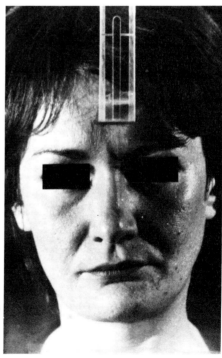

b

Fig. 14.1 a. and b. Photographs and cephalometric films of the head. Profile and frontal view before operation showing the short face with a pronounced sublabial fold, a vertical microgenia, and an incisive over-eruption.

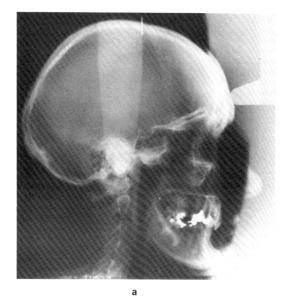

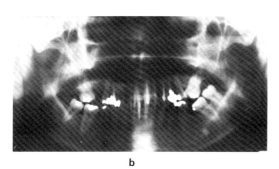

Fig. 14.2 a. and b. Panoramic X-ray of the jaws before the operation showing the incisive over-eruption and the concavity of the submental region.

operation show the different anatomical elements characteristic of the short face.

The patient analyzed here showed a super-eruption of the superior teeth associated with a super-eruption of the inferior teeth, which accentuated the dysmorphosis even further. The concavity of the basilar border was very marked and in profile the basilar border went upwards in front whereas normally it goes downwards in front, thereby outlining the contour of the chin. On the other hand, the basilar border was excessively developed towards the front, giving a chin with a 'galosh' shape. These different morphological elements are very important to note, since this will permit a good surgical correction. The

cephalometric study is particularly important since it provides an appreciation of the volume of the skeletal bone, the contours of the soft tissues and the relationships between them. One often has the impression that the surface of the soft tissues is normal whereas in reality its distribution is disturbed by the dysmorphosis of the bony skeleton.

This brief description makes it possible to understand that operations on the maxilla involving a horizontal osteotomy of the Le Fort 1 type with downward displacement and a bone graft to augment the height of the lower third of the face are often inadequate. It is much more effective to intervene in the area of the mandibular symphysis after correction, if necessary, of the superior incisive super-eruption.

To do this we have devised two types of procedure which can be combined or carried out independently. They permit improvement of the skeleton and the cutaneous contours. We took our inspiration from the Köle procedure to begin with.

The symphyseal region is approached through a vestibulolabial incision 4 cm in length, parallel to the base of the vestibule and approximately 1 cm from it. The incision, which spares the frenulum, extends into the musculature of the mental crest down to the periosteum. This approach makes it possible to leave the muscular insertions in the alveolar region and conserve the vestibule intact. Excellent layers are obtained for closure which are easily sutured. One must be careful about two things:

1. do not perforate the skin, which is often quite close, particularly in the area of the sublabial fold;
2. do not damage the branches of the mental nerves which fan out in this area very superficially.

The subperiosteal dissection goes vertically down to the basilar border which is preserved. Laterally, undermining extends to the mental foramina where the nerves are preserved. Finally the fibro-mucosal layer is undermined in the area of the canines and bicuspids on both the vestibular and the palatal sides. This must be done carefully in order not to tear the fibro-mucosal layer which is relatively fragile. Then the mucosa is tunnelled from the vestibular side to the mental nerves, and from the lingual side down to the line of insertion of the mylohyoid.

Then osteotomies can be carried out.

To correct the super-eruption of the lower incisive teeth, the incisive–canine segment needs to be lowered. To do this, it needs to be mobilized and a horizontal subapical osseous segment resected. The height of the resection depends on the extent of the lowering to be performed (generally 3 to 4 mm). This is done with a bone burr as far back as the inner table in order not to

damage the soft tissues of the floor of the mouth (Fig. 14.3a and b).

The vertical sections (between the canines and the first bicuspids) are carried out with a fine bone burr avoiding the roots of the teeth. If, radiologically, there is very little space between the canines and the first bicuspids, particularly in the cervical region of the tooth, the section of the medullary bone should be carried out with a very thin osteotome. During this part of the operation (often difficult to carry out), one must preserve the

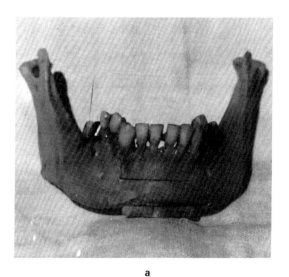

a

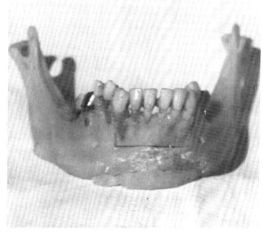

b

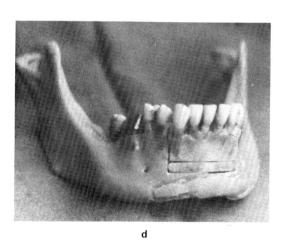

c

d

Fig. 14.3 Photographs of a mandible showing sectioning of the inferior incisive–canine segment. **a.** Cutting the subapical segment (which will be elevated); **b.** cutting an anterior symphyseal osseous segment (which will be removed); **c.** lowering of the incisive–canine segment; **d.** insertion of a segment (subapical) at the level of the lower border to reinforce the vertical chin.

mucosa (which passes as a bridge above the os-teotomy line) because it is this mucosa alone which will assure the blood supply of the bony segment. Using a small periosteal elevator which is placed in the osteotomy line one can gently mobilize the bony fragment to break the remaining bridging portions of bone. Finally, one finishes the horizontal resection on the lingual side which permits the immediate lowering of the dental fragment.

Closure of the soft tissues is done in two layers: a very precise musculo-periosteal layer with approximation of the horizontal planes because the skeleton has undergone an increase in height; a mucosal layer after undermining the submucosa.

The incisive–canine fragment is stabilized with a rigid arch wire leading from the right second molar to the left second molar, passing around the distal surfaces of the teeth to avoid any tilting of the fragment towards the front. This arch wire is fixed to the incisors and the canines with ligatures of 0.3 wire and to the premolars and molars with 0.4 wire. To finish, security wires of 0.3 are placed horizontally between the canines and the first bicuspids. The osteosyntheses, quite loose at the base of the dental fragment, allow its positioning without interfering with the dental arch wire.

At this point of the operation (the time of the repositioning of the dental fragment) the dental occlusion is corrected as completely as possible avoiding any superocclusion.

It remains for the chin to be given a normal vertical volume. To do this, we resect the excess anterior basilar bone (which causes the 'galosh' deformity of the chin). We use a fine cylindrical bone burr which makes it possible to detach the lateral table. This is sectioned vertically for 20 mm on each side of the midline. We preserve the muscular insertions which provide the vascularization of the graft. This (generally of a convex shape) is then placed under the concave basilar border. It is satisfying to note that the contours always fit together which makes their adaptation excellent. To stabilize the graft it is necessary to place several small osteosyntheses of 0.3 wire.

The incisive–canine fragment is stabilized also by two 0.3 osteosynthesis wires.

When it is not necessary to reduce the volume of the chin horizontally (which is rare) we use the subapical fragment as a graft; we place it under the basilar border and fix it as noted above (Fig. 14.3c and d).

A compressive dressing of elastoplast makes it

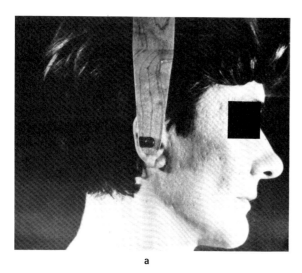

a

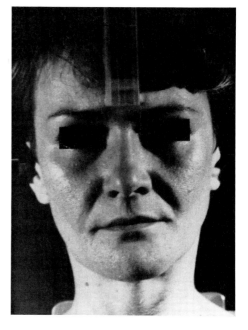

b

Fig. 14.4   a. and **b.** Profile and frontal view after operation showing the lowering of the inferior incisive–canine segment, reinforcement of the chin vertically, disappearance of the sublabial crease and the anterior projection of the chin, and correction of the abnormal height of the lower portion of the face.

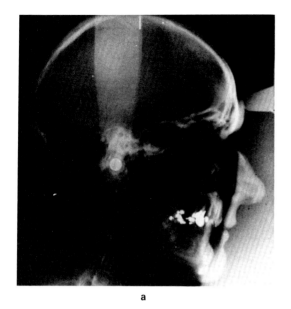

a

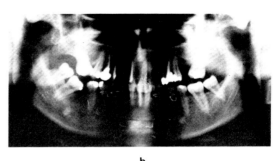

b

Fig. 14.5 a. and b. After operation, showing inferior repositioning of the incisive–canine segment, and reinforcement of the submental region by an anterior cortical bone graft.

possible to provide sufficient pressure on the sublabial region and symphysis to avoid a haematoma and provide good wound healing.

This procedure thus makes it possible to give the lower portion of the face a normal height, reconstitute a vertical chin of a satisfactory volume and correct the height of the dental arches without taking a bone graft from a distant area. In addition, the excess portion of the anterior mental region is removed as required and the labiomental fold disappears immediately.

The harmony of the lower portion of the face is re-established under ideal conditions (See Figs 14.4 and 14.5).

Index